Dissertating Geography

This book examines the history of geography (1950–2020) from a bottom-up perspective.

Disciplinary histories often emphasise the pronouncements of established academics, yet student-geographers make up the majority of the overall 'geographical community' at any one time. Exploring these efforts of geography students over the past 70 years places the known history of the discipline in a new perspective. A disciplinary history 'from below' recognises and acknowledges student dissertations and advances three core propositions: first, they are produced by an overlooked but nonetheless central grouping in the geographical community; second, the rich archival collection of dissertations specifically consulted here contains many excellent geographical knowledge productions that have remained barely read until now; and third, there is a wish to encourage others to explore similar collections of student knowledge productions held elsewhere.

This book will be an important resource for scholars and postgraduate students in Geography, Education, and the History and Theory of Geography.

Mette Bruinsma is an Assistant Professor in the Cultural History research group at Utrecht University, focusing on history of knowledge and university history.

Routledge Research in Historical Geography

This series offers a forum for original and innovative research, exploring a wide range of topics encompassed by the sub-discipline of historical geography and cognate fields in the humanities and social sciences. Titles within the series adopt a global geographical scope and historical studies of geographical issues that are grounded in detailed inquiries of primary source materials. The series also supports historiographical and theoretical overviews, and edited collections of essays on historical-geographical themes. This series is aimed at upper-level undergraduates, research students, and academics.

Earth, Cosmos and Culture
Geographies of Outer Space in Britain, 1900–2020
Oliver Tristan Dunnett

Recalibrating the Quantitative Revolution in Geography
Travels, Networks, Translations
Edited by Ferenc Gyuris, Boris Michel and Katharina Paulus

American Colonial Spaces in the Philippines
Insular Empire
Scott Kirsch

Empire, Gender and Bio-Geography
Charlotte Wheeler-Cuffe and Colonial Burma
Nuala C Johnson

Dissertating Geography
An Inquiry into the Making of Student Geographical Knowledge, 1950-2020
Mette Bruinsma

For more information about this series, please visit: https://www.routledge.com/Routledge-Research-in-Historical-Geography/book-series/RRHGS

Dissertating Geography
An Inquiry into the Making of Student Geographical Knowledge, 1950-2020

Mette Bruinsma

LONDON AND NEW YORK

First published 2024
by Routledge
4 Park Square, Milton Park, Abingdon, Oxon OX14 4RN

and by Routledge
605 Third Avenue, New York, NY 10158

Routledge is an imprint of the Taylor & Francis Group, an informa business

© 2024 Mette Bruinsma

The right of Mette Bruinsma to be identified as author of this work has been asserted in accordance with sections 77 and 78 of the Copyright, Designs and Patents Act 1988.

All rights reserved. No part of this book may be reprinted or reproduced or utilised in any form or by any electronic, mechanical, or other means, now known or hereafter invented, including photocopying and recording, or in any information storage or retrieval system, without permission in writing from the publishers.

Trademark notice: Product or corporate names may be trademarks or registered trademarks, and are used only for identification and explanation without intent to infringe.

British Library Cataloguing-in-Publication Data
A catalogue record for this book is available from the British Library

ISBN: 978-1-032-39039-0 (hbk)
ISBN: 978-1-032-39040-6 (pbk)
ISBN: 978-1-003-34813-9 (ebk)

DOI: 10.4324/9781003348139

Typeset in Times New Roman
by SPi Technologies India Pvt Ltd (Straive)

To Jochem, Olivia & Faas

Contents

List of figures	x
List of tables	xi
About the Author	xii
Acknowledgements	xiii

1 Who actually makes the history of geography?   1
  Student-geographers: producers or consumers of geographical knowledge? 3
  Dissertations as intellectual, cultural, and social source material 4
  The identity, or identities, of geography 6
  Exploring the history of geography through hundreds of geography students 8
  Intermezzo 1: New Towns 11
  Bibliography 14

2 The history of geography and educational practices   15
  The wider context of British higher education 16
  The history and future of dissertations 17
  Dissertations as learning tool or assessment tool 20
  The role of the supervisor 22
  The broader departmental and intellectual network 23
  Increasing explicit governance of the dissertation process 25
  The dissertation and the curriculum 26
  Dissertations and authorship 28
  Conclusion 30
  Intermezzo 2: Outdoor Recreation 31
  Notes 32
  Bibliography 33

viii Contents

3 **Spatial contexts of student knowledge production: the expanded geographical field** 35
   The geographies of undergraduate dissertations 36
     Study areas in Scotland 38
     Study areas in the rest of the UK 39
     International study areas 40
   Emergence of the microscale: biographies, bodies, and bothies 41
   Going into the field 43
   The practicalities of fieldwork 46
   Assistance and help in the field 50
   Field expeditions 52
   Conclusion 55
   Intermezzo 3: Rural Depopulation 56
   Notes 58
   Bibliography 58

4 **Becoming a geographer: dissertations as intellectual source material** 61
   One undergraduate degree with two distinguished pillars 62
   Recent bridges between human and physical geography 63
   Subdisciplinary shifts 67
   Geomorphology: describing landscapes, modelling landscapes, or explaining landscapes? 69
   The explosion of social and cultural geography from the mid-1990s 75
   Conclusion 82
   Intermezzo 4: Social Justice 84
   Notes 85
   Bibliography 85

5 **Geographical traditions versus innovations: students as drivers of disciplinary change** 87
   Disciplinary awareness 87
   The cohort of 1998: disciplinary awareness and conceptual framing 92
   Conceptual frameworks 96
     Regional geography 97
     Spatial Science 99
     Marxist geographies 100
     Humanistic geographies 102
     Other conceptual frameworks 103

Conclusion 106
Intermezzo 5: Medical/Health Geographies 107
Notes 110
Bibliography 110

6 **Exploring the skills of geographers-in-the-making**   **113**
Methods of data collection 113
  Observing 113
  Measuring 117
  Counting 118
  Asking 121
  Mixed methods approach 122
Methods of data analysis 123
  Structuring, categorising, and calculating data 123
  Statistical analysis 123
  Coding 125
  Mapping landscapes versus modelling landscapes 127
  Methods and research design frameworks 131
Ethics and positionality 131
Graphicacy as the geographical skill 134
  Maps 135
  Photographs 135
  Diagrams 140
Conclusion 143
Intermezzo 6: Commuting 144
Notes 145
Bibliography 145

7 **Reflections on student journeys into geography**   **148**
Disciplinary histories from below 149
The lived experiences of geography students 'entering' the disciplinary community 150
Dissertations as sources 152
Closing the cupboard door 155
Note 156
Bibliography 156

Index   *157*

Figures

1.1	New Town research polaroids (Scullion, 1974)	12
2.1	Map of 'caravanning' capability, based on slope, texture, and drainage (Jarvie, 1974)	32
3.1	Photo of the flume, located at the School of Engineering (Baff, 1994: 17)	45
3.2	Dissertation cover (Gallacher, 2002)	46
3.3	Land use survey photographs (Waddell, 1974: n.p.)	47
3.4	Cover of dissertation based on field expedition to Tanzania (Scholes, 2010)	54
3.5	Graphs of questionnaire results, displaying 'major reasons why people want to leave Islay' (McMinn, 1982)	57
4.1	Number of dissertations per cohort that can be categorised as human geography, physical geography, or 'both'	63
4.2	Percentage of environmental geography dissertations per cohort	64
4.3	Number of dissertations per subdiscipline	68
4.4	Percentage of glacial geomorphology dissertations per cohort	71
4.5	Number of social geography dissertations	76
5.1	Dissertation cover (Todd, 1990)	91
5.2	Dissertation cover (Kitchingham, 1998)	95
5.3	Map of free meal uptake in Lanarkshire schools (Murphy, 1994)	108
6.1	Landscape sketch: Loch Riddon (McAllister, 1966)	115
6.2	Codes and examples (Fleming, 2010)	127
6.3	Example of coding process (Adamson, 2010)	128
6.4	A map of suggested ice movement (Young, 1978)	130
6.5	Polaroid (Crabb, 1958)	136
6.6	Dissertation cover (Birch, 1986)	137
6.7	Semi-diagrammatic sketches (McAllister, 1966)	139
6.8	Fieldwork sketches of Airdrie (Kelly, 1974)	140
6.9	Norse house types, observed in the field (Burns, 1966)	141

Tables

2.1 Prescribed books for academic year 1954–1955 (further readings recommended during the course) (University Calendar, 1954–1955, SEN10/97)   27
3.1 Number of dissertations per study area category   37
3.2 Number of dissertations per study area within the UK (but outside of Scotland)   39
3.3 Top-10 of countries (outside of the UK) most studied   40
3.4 Overview of other countries or areas studied at least once   41
4.1 Subdisciplines: cohort in which they were most popular   69

Author

Dr Mette Bruinsma, having completed a MA degree in the History and Philosophy of Science, worked as an educational policy maker at the Faculty of Geosciences at Utrecht University. Here, she became fascinated by geography's 'struggles' concerning disciplinary identity and the diversity in methods and concepts used by the discipline. In 2017, she moved to Glasgow, UK, to start a PhD project funded by The Leverhulme Trust. Her PhD focused on undergraduate geography education and geography's disciplinary history. In 2021, Mette passed her viva, moved back to the Netherlands, and pursued her academic career in the Cultural History research group at Utrecht University as an assistant professor, focusing on history of knowledge and university history.

Acknowledgements

I am very grateful for *The Leverhulme Trust*'s funding for this project.

I would also like to thank the two great people who supervised my PhD research, which formed the basis for this book. *Chris Philo*, thank you so much for your non-stop enthusiasm for this project, your incredible memory of every single undergraduate dissertation you ever supervised or marked, and all your knowledge and encouragement. *Cheryl McGeachan*, you got involved in this project halfway through: you really brought something new to the team, and your non-stop enthusiasm was contagious. It was so good to have to explain my research to someone who wasn't there from the start. Thank you so much for stepping in the way you did.

Thanks to all my *colleagues and students at the Cultural History department at Utrecht University*. I am incredibly lucky to work in such an inspiring and stimulating environment.

Thanks to all the *interview participants*, and all *Glasgow's geography dissertation writers*: reading about all the dissertation joys, struggles, travels, knowledge productions, and adventures was so inspiring and fun!

And most of all, thanks to *Jochem*, *Olivia* and *Faas*: forgive me for the innumerable times I will start my sentences with "Remember, when we lived in Glasgow..." in all the years to come. I am the luckiest person to be with the three of you, and to have had this great adventure with you in Scotland while pursuing this research project. Dankjulliewel alledrie, voor alles.

1 Who actually makes the history of geography?

In 2019, a group of former geography students met up at their *alma mater*, in Glasgow, Scotland, for the reunion of 'the Class of '94'. The alumni present at the reunion exchanged stories of their lives after graduation – careers, houses, health, and children – but there was also a clear vibe of nostalgia about their time as young geographers. When professor Jo Sharp, part of the welcoming party of this reunion, announced that the 70 undergraduate dissertations of Class of '94 were dug up in the archive, a loud sigh followed by a collective 'Oh nooooo!' could be heard in the lecture theatre. Nostalgia, then, was suddenly mixed with feelings of pride, shame, and curiosity about one's own former work. Stories were shared about how much (or sometimes how little) they remembered from these 'dissertation-days'. For some, the dissertation was a starting point of a life-long interest and career in a certain direction, whether that was housing, environmental issues, or urban public transport. Some others expressed feelings of regret: 'why did I think I wasn't good enough to continue with this research? Looking back, I think my dissertation is actually quite a nice piece of work!' Twenty-five years before, this group, a cohort of 70 students, produced these undergraduate dissertations – small geographical knowledge productions, some exciting, innovative, and well-written and others perhaps slightly boring and predictable. These dissertations were read by a supervisor, a second marker, and perhaps the student's parents and maybe a friend or two. But for most of the dissertations: that was it. They were stored, archived, and forgotten.

In this book, the voices of many geographers who are often neglected – student-geographers – in disciplinary histories are staged and centralised in this narrative on the roots, identity, traditions, and innovations of geography as an academic discipline. Although the knowledge produced by geography students is usually seen as ephemeral source material, read by just a handful of people, it is exactly this knowledge that is central in the work happening within the walls of university departments: not just in number of people but also regarding the substantial workload of staff supervision. Writing a dissertation is a shared, formative experience for many, if not all, established academics. The social and intellectual roles that students play in a department are significant, as are the social networks and connections that 'take place' beyond the words written down in the dissertation. The significant experience of writing a dissertation, undertaken by so many, also makes the dissertation a historical source that opens a window on an important rite of passage: annually done by so many, almost a ritualistic movement from being 'just' a student-geographer to being, as it were, an approved, acknowledged, even 'certified' geographer with the degree and academic title as 'proof'. These rituals shift slightly over time, but are nevertheless still very much recognisable as a shared experience across time.

Social networks and socio-economic backgrounds of students raise unequal opportunities and the facilities and support within the university are also different for every university, every cohort, and every individual student. Societies change, universities change, curricula change and students change, and all these changes mean that critical historical analysis of how geographical knowledge has been produced by many generations of students must evoke many specific questions about the circumstances in which this knowledge has been produced. The central role of 'fieldwork' in the geography degree adds a specific additional flavour to these questions, namely the status of independent research away from university, causing a spatial distance between a student and university. The geographies of undergraduate students disclose worlds unto themselves: from laboratory to library, from rivers to mountain tops, and from chicken coops to shopping centres, student-geographers go just about anywhere, usually physically but sometimes only virtually. In work on the history of academic education, the 'student experience' of undergraduate students is regularly studied, but the first individual *research* endeavours of undergraduate students are not; witness the almost complete absence of extant work on the history of 'the dissertation' in comparison to research on, for instance, doctoral theses (e.g. Kaplan and Mapes, 2015). However, these experiences turn out to be highly formative experiences for many future researchers and play a similar role in the lives of geographers with other career paths. The lively and detailed memories that people share when asked about their undergraduate dissertation research prove that this element of the curriculum is something different from regular coursework.

The choices made by many generations of undergraduate students reveal longitudinal trends and shifts in the discipline and its component subdisciplines, but they do offer more than that: in discussions about canonicity, or the lack of a geographical canon, student-geographers as a wider grouping within (or at least fringing) the academic community prompt alternative ways to look at canonicity. The bibliographies of the dissertations disclose transient 'temporary canons' that are widely shared by students at a specific moment and a specific place in time: for instance, the usage of the Statistical Accounts was ubiquitous in the regional dissertations of students of the 1960s, as was Flowerdew and Martin's *Methods in Human Geography* (2005) for more recent human geography students around 2010. These temporary canons seem strongly shaped by members of staff. Sources used by undergraduate students will presumably influence later reflections on 'the geographical canon' when some of the undergraduate students become early-career researchers. This book draws attention to the insights that dissertations offer for the history of geography, but also addresses the complex and hybrid spaces in which geographical knowledge is produced, acknowledging the shared – maybe momentarily or even merely locally 'canonical' – foundations of future 'established' academic geographers as well as of many 'professional' geographers continuing their careers outside of the university.

Student-geographers: producers or consumers of geographical knowledge?

Geography as a discipline is practised by a relatively small group of academics, but is also practised by a much greater number of students, these numerous apprentice geographers, every day. Students, in their roles of geographers-in-the-making, get to know the discipline through textbooks, besides, of course, from the direct influence exerted by academic staff members within their department (Sidaway and Hall, 2018). Research on the history of geography with textbooks as a starting point inherently position the geography students using these textbooks as merely *consumers* of geographical knowledge. There are also, however, examples of studies that seriously consider geography students as independent geographical knowledge *producers*. Examples of giving a voice to students in the history of geography can be found in an account of the student-led annual journal *Drumlin* of the Glasgow Geography Department (Philo, 1998), and also in the 'restorying' of a 1951 school field trip to the Glenmore National Park (Lorimer, 2003). Philo's analysis of the journal pays attention to how student-geographers receive, respond to, and perhaps reject the geographical knowledge presented to them by their academic teachers. He argues that, in contrast to what occurs in most other historiographies, not only the production of geographical knowledge by established academics should be studied but also its consumption and then *re*production by students. Addressing earlier literature on the geographical knowledge of students, Philo states:

> Yet little appears to have been said about how academic visions of geography have been received by the countless students who filter through 'our' lecture theatres, tutorial rooms and libraries. It seems to me that this is an omission of some magnitude, and that more might be done to ascertain how all manner of shifts in academic geography, as fostered by small numbers of 'visible' professional geographers, translate into the thinking, writing, fieldworking and murmurings of the many 'invisible' student geographers whose responses are almost never afforded sympathetic scrutiny.
>
> (1998: 345)

This quotation displays three distinct elements that are also essential to this book. First, the 'countlessness' of students: the fundamental viewpoint that in studying the history of a specific discipline, the greatest number of 'practitioners' of this discipline should not be overlooked. Students graduating from a geography undergraduate degree vastly outnumber professional academics in geography, and these many students are nonetheless still surely, very legitimately, called 'geographers'. Second, the spaces in which this transaction or exchange of knowledge happens, including these spaces within geography departments or school buildings. Third and last, the 'academic visions of geography that have been received' are arguably even more fully digested and considered in the choice and execution of every student's independent research project: the undergraduate dissertation. As Philo addresses, adding the student knowledge productions to the narratives about the history of geography does not necessarily offer 'entirely surprising new species of geographical knowledge' (Philo, 1998: 361), yet the emphasis on the small knowledge productions in the undergraduate dissertations is nonetheless exactly that: knowledge productions. Some of these do add something innovative or surprising to the discipline. Others are merely 'small stories' (Lorimer, 2003) in themselves: adding institutional, intellectual, personal, and educational perspectives on what it means to be a geographer-in-the-making. Encountering these knowledge productions of students in the historiography of geography touches on the intellectual and social experience of every student: the process of coming up with a research topic, doing independent research, and finally the writing-up of the dissertation. It is a fundamental and formative process in becoming-a-geographer, and both these practices and the actual knowledge produced in these undergraduate dissertations are valuable but overlooked elements in the historiography of geography (Bruinsma, 2021).

Dissertations as intellectual, cultural, and social source material

Disciplinary histories of geography often emphasise the works of established academic geographers. Such works are often envisaged as taking place in grand scholarly spaces, such as monographs, academic journals, and conferences. Students who complete their geography undergraduate degree studies vastly outnumber 'professional' academic geographers. However, their works normally remain invisible, unconsidered, and seemingly of no account in the making of academic geography or, more broadly, geographical knowledge. Of course, their works – their essays, lab work, group assignments, and other products – are for the most part ephemeral, destroyed, or deleted by their institutions after a given time period. The only works that might escape this fate are the so-called dissertations, which sometimes remain stored for longer, perhaps becoming a nuisance clogging up departmental cupboards or filing cabinets. This book takes such a store or 'archive' of dissertations as its point of departure, allowing to ask questions about how disciplinary histories of geography might be supplemented, and perhaps in some respects changed, by focusing directly on these otherwise forgotten works by many novice or apprentice – but still, as is argued here, academic – geographers.

The case study explored in this book includes the dissertations produced by many generations of undergraduate geography students of one British university: the University of Glasgow. The geography undergraduate dissertation archive, comprising over 2,600 dissertations from the early 1950s to the present (the dissertation collection is growing year by year, with new students 'inhabiting' the department and writing their dissertations as fourth-year students), offers the opportunity to conduct a longitudinal study of formally similar – they are all undergraduate geography dissertations – but actually, in practice, highly variable sources. The word 'dissertation' is used here for all these sources, even though the documents themselves were not consistently named 'dissertations' until 1975. These changes in vocabulary hint at changes in the curriculum, rules and regulations, and expectations of what a 'dissertation' should encompass. This archive's size and scale bring distinctive opportunities for historical research, allowing sound comparisons to be made across time, theme, cohort, and more, as apparent from the chapters that follow.

An archival collection of such dissertations arguably encompasses three kinds of sources: *intellectual sources*, because every dissertation includes an original piece of academic research, however modest in concept and execution; *cultural sources*, because the dissertations are often experienced as rites of passage in the process of moving from being an undergraduate geography student to being a 'real' qualified geographer; and, lastly, as *social sources*, whereby the dissertations illuminate diverse personal networks of peers, family, supervisors, other departmental staff members, research participants, and external collaborating individuals and organisations. This book addresses the value of exploring extensive collections of dissertations from a disciplinary historian's perspective. Such collections may be researched both as an archive full of individual sources – particular dissertations that ideally should each be read from cover to cover – and also as a singular source, one collection, as a whole – a source that is the quantitative sum of all these 'small' knowledge productions. Acknowledging the epistemological value of these small knowledge productions, as well as the formative and recognisable experience of becoming-a-geographer, enriches the history that can be told of academic geography.

These undergraduate dissertations are not entirely 'free forms' of writing, since there are indeed many rules and regulations with which a student needs to comply in order to be able to produce a successful and acceptable dissertation. Power relations – between supervisory staff and student, between university and student, and perhaps between expectations of parents and their studying adult children – are central to this inquiry into geography dissertations. It is in the cross-pollination of concepts and ideas from the history of geography, the history of academic education, and the geography and sociology of scientific knowledge production that the coordinates for this research are therefore located. Knowledge productions in undergraduate students' dissertations obviously hold potential for pedagogical inquiry, but these novice voices also speak of disciplinary trends, traditions, and innovations within a complex, hybrid, and situated social space. The dissertation archive bears witness to changes in how students were instructed about the dissertation process, about what exactly they could expect from supervisors (and vice versa), and how 'dictated' has been the structure of the dissertation itself. The 'small voices' of the ostensible authors – the students – speak of innovative, or sometimes not so innovative, research projects: small in their scope, but even smaller in their audience. Not all these knowledge productions perhaps add a new view to our discipline, but studying a wider collection of the dissertations surely does tell us something fresh as a potential new perspective on disciplinary histories, traditions, cultures, sociologies, and more.

The size of the Glasgow collection of undergraduate geography dissertations and the opportunities offered for a longitudinal study of these similar yet distinctive sources also enable the recognition of notable overall trends and breaks. The archival collection full of dissertations comprises a collection of personal memories, intellectual knowledge productions, and clues in shifts and changes in what it meant to be a geographer-in-the-making. Whereas geography is often seen as a discipline that is difficult to define, these shifts – as well as some striking continuities, such as the practical challenges encountered while doing fieldwork and the use of visual materials such as photos to support the research – demonstrate changes in dominant conceptions and expectations of what a geographer should know, what skills a geographer should possess, and what kind of questions one should ask: changes occurring sometimes abruptly and other times slowly and almost silently. Studying geography is a shared basis for becoming-a-geographer. Based on the undergraduate dissertation archives of Glasgow Geography, this book hence provides a bottom-up, student perspective on the history of geography. With these geographers-in-the-making and their knowledge productions as a starting point, the existing historiographies of geography are extended with these 'voices from below'.

The identity, or identities, of geography

The history of geography is in part a history about a discipline struggling to define its unity and, related to that, its identity. This struggle is not a purely intrinsic one with, or indeed between, the distinguishable pillars of human geography and physical geography. Even more than other academic disciplines, perhaps, geography is not isolated. This is perceptible in the administrative place of geography in British universities. Geography degrees regularly share administrative units in universities with disciplines such as environmental science, sociology, and geology, but also with more surprising disciplines: for instance, chemistry and design and technology (Hall et al., 2015: 61). This administrative diversity in where geography is located within academic institutions mirrors the hybridity of interdisciplinary roots and connections carried by geographers, in terms of research collaborations, shared methodologies, and conceptual and theoretical frameworks. Although the caricature of 'camps' between human geographers on the one hand and physical geographers on the other, as sketched by Inkpen (2018), is recognisable within institutions and from the relative presence or lack of historiographical literature offered by the different camps, some geographers have actively tried to search for the unifying elements within geography, such as Matthews and Herbert (2004). This hybridity of the discipline can be seen as a problematising factor in terms of fixing 'the' identity of the discipline among other disciplines. Others are not looking for a unification of geography, however, but are embracing the heterogeneity and seeing it as potentially beneficial (Kwan, 2004). The theme of hybridity and unity within geography and geography's relationship with other disciplinary traditions returns in the analysis of the undergraduate dissertation archival collection: not only on a microlevel – how particular students reflect upon the discipline of which they are effectively members – but also on a macrolevel – how the archival collection as a whole might reveal a changing disciplinary awareness and ambition over time. The undergraduate degree programme that comprises human and physical geography might even be cast as the unexpected primary unifying practice that makes geography one discipline. Putting a central focus on both students' experiences and students' knowledge productions in this bottom-up narrative of the history of geography hence provides an opportunity to approach geography as a hybrid but unified discipline.

The undergraduate geography degree in the UK, as it is taught in many British universities, distinguishes itself from earth sciences, but does combine human, physical, and environmental geography in the undergraduate curriculum. This geographical disciplinary and educational 'unity' unavoidably evokes questions about the historical and philosophical identity of the discipline (Van Meeteren and Sidaway, 2020). Although this book argues for the potential value of undergraduate dissertations as sources of disciplinary history, as well as being potentially innovative and insightful sources for contemporary research in a variety of disciplines, dissertations are particularly telling in respect of *geography*'s disciplinary history and identity. The adjectives to describe geography are complex, diverse, and sometimes a 'bit much':

> Geography is a sprawling, ragged, gorgeous, discipline. It ranges across the physical and social sciences into the humanities and the performance arts. It's a discipline with a whole heap of different ways of doing what it does. It maps and models. Critiques and exposes. Drills and digs. Surveys and measures. Talks and hangs out with. Theorises. Analyses. Deconstructs. It's a discipline that both knows what it's about, and yet were you to ask a group of academic geographers what exactly it is that defines geography each would give a different answer. Stuffy and hip, it's a discipline with too much difference for some and yet not nearly enough for others.
>
> (Geoghegan et al., 2020: 462)

This book explores such questions on 'the geographical identity' and also on how the (self-) positioning of geography and of geographers become perceptible in the longitudinal study of the collection of undergraduate dissertations. This internal debate within geography about what geography actually *is*, or what it *should be*, makes the discipline particularly interesting for this exercise of including the many new voices appearing within the discipline over time: the voices of geography students, who, at some point, decide that geography (following their interpretation of what geography actually entails) is really what they want to do.

The roots of geography are tangled in many disciplinary traditions but, above all, the discipline has been strongly influenced by convergences to and divergences from the natural sciences. This historically recurring emphasis on the kinship of geography to the natural sciences underlies the identity struggle of geography: to what extent is geography actually one discipline if its roots and ambitions are scattered? All the different ways to try to grasp the discipline, finding some kind of core or unity, and any way of structuring geography's past, have their own limitations. In the search for this core, the identity of the discipline is troubled, because geography does not have one or even a few central concepts or practices, but rather it has many (Clifford et al., 2009). Trying to 'fit' geography into the discourse of natural laws and emphasising quantitative methods has often been undertaken to 'reinforce ... scientific authority' (Livingstone, 1992: 326). The starting point of this book lies in the 1950s: a decade when academic geography was strongly influenced by the classic 'scientific method', as well as by the lingering appeal of the more descriptive practices integral to regional geography. In later chapters, these influences and traditions will be discussed further. The experiences and knowledge productions of novice geographers provide a unique perspective on such questions. Authors of the dissertations position themselves on various perches in the shifting landscape of disciplinary allegiances to natural sciences, social sciences, and, more recently, the arts and humanities. Considerations concerning the status and identity of the discipline are implicitly, and sometimes explicitly, expressed by many undergraduate students. Their reflections, yielding examples of students 'self-positioning' their work within the discipline, thereby enrich the traces and intellectual kinships that can be found within the history of geography.

Exploring the history of geography through hundreds of geography students

Before turning to the dissertations themselves, in Chapter 2 the interconnectedness of the history of geography and educational practice is put central to the discussion. Undergraduate dissertation research is a distinctive form of educational practice. Over time, the nature of this entity called 'the dissertation' has changed – with particular national and local developments and discussions influencing such changes. This chapter will offer some further perspective on the 'weight' of the dissertations for students as well as for supervisors and markers. Chapter 2 will address the position of the undergraduate dissertation in British higher education in general, as well as the local 'varieties' of the relationship between dissertations and undergraduate geography curricula.

Spaces and contexts of knowledge production are, of course, central in the research geographers *do*, but such spatialities are also vital to anyone interested in the production of geographical knowledge *itself*. Chapter 3 considers these social and spatial contexts of geographical knowledge production. The university department may be seen as a social space of knowledge production, in which professional, established academics do their research. However, the same can be said about the many past and present cohorts of geography *students*, who also 'inhabit' the department for a limited period of time. For instance, issues of globalisation are tracked in relation to the affordability of travelling abroad as well as the internationalisation of higher education, perceptible both in an increasingly diverse student population in the department and in more undergraduate students going abroad as part of exchange programmes such as ERASMUS, all traceable in the dissertation archive. Connected to the whereabouts of study areas are also more conceptual discussions concerning the conception of what 'the field' actually comprises, thereby reassessing themes extensively discussed elsewhere by academic geographers. In this chapter, the practicalities of fieldwork and the role of collective, organised field expeditions in the dissertation experiences of many past students are addressed. These discussions further demonstrate the role of wider social networks in doing fieldwork and other dissertation work: family, friends, and fellow geography students have often played the role of 'unpaid research assistants'. By reconstructing the changing practices of dissertation research and writing, a different light can be shed on theoretical and conceptual matters concerning the 'expanded field' as discussed in the discipline's academic and educational literatures.

In Chapter 4, disciplinary shifts and continuities are explored, deploying both quantitative analysis and qualitative interpretation. This chapter comprises analysis of both the intellectual content of one dissertation archive as a whole and of numerous sampled dissertations in particular. First, the relationship between human geography and physical geography is addressed, as well as identifying both older and more recent bridges between the two: for instance, contrasting the area-based syntheses of natural and cultural landscapes common in older dissertations with hybrid studies on sustainability, climate change, and green energy appearing in more recent versions. Second, this chapter recovers subdisciplinary shifts and trends over time, in which respect the decline of regional geography and the rise of the systematic, or subdisciplinary, geographies are clearly visible in the Glasgow case study. In this chapter, causes for the rapid rise – and sometimes fall – of certain popular subdisciplines are discussed, focusing particularly on changes in geomorphological studies over time and the explosion of cultural and social geography studies towards the end of the twentieth century. Broader reflections are offered about how dissertations mirror, anticipate, lag, or simply look different from the standard portrayal of disciplinary history.

10 *Who actually makes the history of geography?*

Chapter 5 is to some extent a continuation of the themes discussed in the previous chapter, but in this chapter it is not about the changing foci of inquiry, but rather students' reflections on both the discipline itself and shifts in its conceptual and theoretical frameworks. The chapter thus addresses two different themes, the first being the disciplinary awareness of students, exploring how students have positioned their own research in a wider geographical disciplinary framework. Both the presence or absence of such disciplinary reflections and the specific ways in which these reflections are written up disclose something about how geography has been taught and how disciplinary reflection, introspection, and self-critique have changed over time. The second aspect are the conceptual frameworks that students have adopted, more or less explicitly, and then used as a justification for what follows in their inquiries. A comparison of the timeline that an analysis of undergraduate dissertations provides with the existing historiographical timelines of conceptual and philosophical shifts within academic geography offers insights into the interactions and relations between professional geographical research and geographical academic education. It also addresses the 'routes' taken by geographers-in-the-making into the discipline of which they are part, as well as showing how, along the way, they may change, reshape, and challenge that same discipline.

In Chapter 6 students' skills are explored, emphasising methods of data collection, data analysis, and skills to display, present, and represent findings. The methods of data collection, such as questionnaires, interviews, participant observation, and surveys, are inexplicably connected to the foci of inquiry and the conceptual frameworks discussed in previous chapters. With changing conceptions of the 'geographical field', there are also shifts in skills needed to collect data in the specific field student-researchers are operating in. The longitudinal nature of this research means that numerous technical developments over time are captured in the archive, influencing the devices used as well as the field and information technology (IT) skills needed to work with (or without!) such devices. The chapter continues with a discussion about the written and visual forms of representation in the dissertation. The bound documents represent endless days spent in the library or elsewhere writing, typing, drawing, and/or mapping. Literacy and 'graphicacy' (Balchin, 1970) are considered in this chapter, paying specific attention to the mapmaking and drawing skills of the many generations of students.

The final chapter, Chapter 7, reflects on the value of recovering the otherwise under-represented voices of student-geographers for disciplinary historians, as well as for other geographers interested in the traditions of their discipline. The chapter explores the main findings, relating to curriculum design, bridges between human and physical geography, and the changing balances between regional and subdisciplinary inquiry, and more; as well as the potential for further exploring student knowledge productions, such as the use of quantitative and qualitative methods to explore the contexts and contents of undergraduate dissertations. In this chapter, the increase in explicit self-awareness and disciplinary awareness of novice geographers is centralised, which also calls for a next step: awareness from readers that knowledge produced by students may actually be valuable geographical contributions *in their own right* and not merely as vehicles for training them in the ways of geographical inquiry, critical thought, and transferable skills.

Intermezzo 1: New Towns

British New Town policies after World War Two were related to the problems of urban housing. In 1946, the New Towns Act was passed, directly followed by the foundation of 27 New Towns, of which a few were located in Scotland: for instance, East Kilbride, Cumbernauld, and Irvine. The New Towns policies are regularly discussed in dissertations, from the 1970s to the 1990s, especially applied to specific New Towns in Scotland. The urban geography interests of staff member Ronan Paddison were central in this respect, and there is more that could be added about his impact on the urban but also political geography foci of many dissertations from the 1970s. In 1974, James Scullion wrote his dissertation *Irvine New Town: Population breakdown and social area analyses*.

In this dissertation, Scullion investigated the differences between neighbourhoods within Irvine, and analysed the demographics of the town. He made use of a very basic questionnaire alongside observational work reinforced by photographs. In the questionnaire, he asked questions about the type of housing provided, how long people have lived there, where they lived before, how many people live in the household, and where they work. The questionnaire data led into a quantitative data analysis:

> We have now examined the population of Irvine New Town in some detail: we have broken it down into its constituent parts and examined them with respect to age, sex, family size, origin, reason for migration and mobility; we have analysed the distribution and standard of life of this populations. These were the basis objects of this study and to an extent these objects have been achieved.
>
> (Scullion, 1974: n.p.)

Scullion's dissertation demonstrated how quantitative methods were central to many systematic dissertations of the 1970s, but also how the New Towns policy was a fascinating focus of inquiry for people interested in urban planning and housing. Throughout the 1970s, every year there was at least one dissertation about New Towns (mostly about Irvine or East Kilbride). In the 1980s, the New Towns still gained attention, but the research questions asked changed somewhat: the key concept became 'self-containment', as in 'to what extent do New Towns have all the facilities that their inhabitants need?' J.G. Main (1982) examined this issue in his or her dissertation entitled *The self-containment factor of Britain's New Towns*.

12 *Who actually makes the history of geography?*

Figure 1.1 New Town research Polaroids (Scullion, 1974).

All the new towns have been developed with the ultimate aim of creating self-contained and balanced communities for work and living. ... Yet despite this common goal, some of the new towns have been more successful than others. This ... raises the question as to whether such an 'ideal' state is in fact possible.

Thus, with this in mind, it was decided, for the purpose of this study, to isolate one of the new towns' objectives – self-containment – and, by studying an existing new town, consider exactly what it involves, the problems which may be encountered and subsequently its chances of success.

(Main, 1982: 1)

Main duly confronted the self-containment factor, and in this dissertation, the study area is Cumbernauld, another Scottish New Town. In the conclusion,

Main wrote about the different facilities, or sometimes lack of facilities, and took an interesting gendered approach which, unfortunately, was not unpacked further:

> ... the overall impression of Cumbernauld is that despite the absence of a general hospital, the population is fairly well endowed by community facilities. Indeed with respect to education, Cumbernauld appears to have satisfied the vast majority of its residents. However, the apparent lack of day nurseries may be a cause for concern. Even in a time of equal rights for women, it is still regarded as being a woman's role to look after the children. Therefore, the lack of facilities may, in fact, discriminate against women who may want, or need, to work during normal working hours.
> (Main, 1982: 58)

Unlike in the quantitative study of Scullion, written in 1974, Main used the collected data to make some more critical, engaging statements about the 'self-containment factor' of Cumbernauld. This difference arguably indicates a slow move towards more ethical-political engagement in the dissertations, traceable across the decades.

From quantitative demographic studies to studies of the facilities, to questions concerning well-being and community spirit, by the early 1990s the New Towns were still being researched, but again the emphasis had shifted. In her dissertation *East Kilbride: Has it created communities by the planned methods of social balance and self-containment?* (1994), Janice Murray explored the 'community spirit' in one New Town:

> The central thesis of this study is to determine the social structure and community identity within the neighbourhoods of East Kilbride as initially the New Town policy emphasised social balance and self-containment as a means of encouraging social cohesion at both the neighbourhood and the town level. Through this study it is hoped to discover whether this policy is still important to residents and, if it is still important, whether social aims have been met by direct policy or indirectly – e.g. sorted neighbourhoods through socio-economic segregation.
> (Murray, 1994: 6)

Murray also uses questionnaires and statistical methods, but also mentions 'personal interview techniques' (Murray, 1994: 15) to gain more input. While ostensibly perhaps not *so* different from the Main study, the emphasis here had slid into something that might be deemed more humanistic, being concerned with the 'structures of feeling' held by research participants – their feelings about more intangible features of a place or community, such as its 'spirit' or influence on 'well-being' – rather than more objective measures of population and facility use.

Bibliography

Balchin, W.G.V., 1970. *Geography: An Outline for the Intending Student*. London: Routledge & K. Paul.

Bruinsma, M., 2021. 'The geographers in the cupboard: Narrating the history of Geography using undergraduate dissertations'. *Area*, 53(1), 67–75.

Clifford, N.J., Holloway, S.L., Rice, S.P., Valentine, G., (eds) 2009. *Key Concepts in Geography* (2nd ed.). London: SAGE.

Flowerdew, R., Martin, D., (eds) 2005. *Methods in Human Geography: A Guide For Students Doing A Research Project* (2nd ed.). Harlow: Pearson.

Geoghegan, H., Hall, S.M., Latham, A. Leyland, J., 2020. 'Continuing conversations: Reflections on the role and future of *Area* from the new editorial team'. *Area*, 52(3), 462–463.

Hall, T., Toms, P., McGuinness, M., Parker, C., Roberts, N. 2015. 'The changing administrative place of Geography in UK higher education: Where's the geography department?'. *Area*, 47(1), 56–64.

Inkpen, R., 2018. 'The "smugness" of geographers: Dismantling the caricatures of philosophies in Human and Physical Geography'. *Area*, 50(1), 46–49.

Kaplan, D.H., Mapes, J.E., 2015. 'Panoptic geographies: An examination of all U.S. Geographic dissertations'. *Geographical Review*, 105(1), 20–40.

Kwan, M., 2004. 'Beyond difference: From canonical geography to hybrid geographies'. *Annals of the Association of American Geographers*, 94(4), 756–763.Livingstone, D.N., 1992. *The Geographical Tradition: Episodes in the History of a Contested Enterprise*. Oxford: Blackwell.

Lorimer, H., 2003. 'Telling small stories: Spaces of knowledge and the practice of geography'. *Transactions of the Institute of British Geographers*, 28(2), 197–217.

Main, J.G., 1982. *The self-containment factor of Britain's New Towns*. Undergraduate Dissertation, University of Glasgow.

Matthews, J.A., Herbert, D.T. (eds), 2004. *Unifying Geography: Common Heritage, Shared Future*. London: Routledge.

Murray, J.M., 1994. *East Kilbride: Has it created communities by the planned methods of social balance and self-containment?* Undergraduate Dissertation, University of Glasgow.

van Meeteren, M., Sidaway, J.D., 2020. History of Geography. In: Kobayashi, A. (ed.), *International Encyclopedia of Human Geography* (2nd ed., vol. 7), London: Elsevier, 37–44.

Pate, S.J., 1974. *Delimitation and analysis of retail spheres of influence in southern Lanarkshire*. Undergraduate Dissertation, University of Glasgow.

Philo, C., 1998. 'Reading *Drumlin*: Academic geography and a student geographical magazine'. *Progress in Human Geography*, 22(3), 344–367.

Scullion, J., 1974. *Irvine New Town: Population breakdown and social area analyses*. Undergraduate Dissertation, University of Glasgow.

Sidaway, J., Hall, T., 2018. 'Geography textbooks, pedagogy and disciplinary traditions'. *Area*, 50(1), 34–42.

2 The history of geography and educational practices

Historians and sociologists of knowledge production often emphasise specific sites in which those knowledges are produced. Knowledge, including scientific knowledge, is produced *somewhere* (Haraway, 1991), and probably these sites in which knowledge is produced directly influence the 'content' of these knowledge productions. As Withers (2009: 653) states: 'science ... is produced through place'. The university department might be an obvious place to look at when studying the whereabouts of geographical knowledge production: looking at papers in academic journals, monographs, and conferences, the affiliation of researchers is always mentioned, implying that this location indeed matters in some way. It becomes part of the identity, or at least status or image, of a researcher, because every academic institution evokes certain ideas about 'what kind' of research is undertaken there. A department is an example of an 'everyday site' (Lorimer and Spedding, 2002) of knowledge production. The department broadens the potential scope of those whose voices could or should be included in the histories, sociologies, and geographies of a discipline. It is a place where these 'big names' – the established academic geographers – do much of their salaried work, often emphasised in narratives about a certain discipline, but there are far more 'small names', as it were, associated with such places. There is, then, a lot of potential for researching departments as 'everyday places' with the 'everyday people' in it: for instance, the many student geographers-in-the-making.

This book addresses the knowledge produced by undergraduate students at The School of Geographical and Earth Sciences at the University of Glasgow: one changing department, a changing curriculum, and many different students and staff members 'inhabiting' this physical and social space. Changes in the discipline are not just about intellectual changes, but also about changes in the people doing the research and changes within institutions (Hall et al., 2015: 56). Formally, teaching and researching geography in Glasgow started in 1909 (Philo et al., 2009). Things that could be called 'Geography' were taught before 1909, but in 1909, the first officially titled lecturer in geography was appointed. A 'story from below' combines some of the questions asked within the domains of the fields of history of geography, history of academic education, and the sociology and geography of knowledge production. First of all, however, it is important to address some key observations about the role of the dissertation as 'key' element of undergraduate education in the British context, before shifting from the national context to one of many British research universities, to see how knowledge production situated in a specific local context affects geography as an academic discipline.

The wider context of British higher education

After 1992, there was a significant increase in higher education institutions (HEIs) prompted by the Further and Higher Education Act. Simultaneously, the total number of students enrolled in these institutions increased as well. The rapid growth of the number of HEIs led to pressure on quality assurance and accountability (Brown, 2004): how would it be possible to maintain the desired level of quality of both research and education, with so many different institutions? The 'apparent variability' of standards of assessment practised by individuals, departments, institutions, and disciplinary communities (Chapman, 1994, in Webster et al., 2000: 3) was a political concern, both recognised within and outside of academia. Before, with few universities and lower student numbers, the number of degree qualifications was relatively small and it was thought possible to ensure comparability and equality in academic standards (Pepper et al., 2001: 25). The traditional form of regulation was 'self-regulation' (Brown, 2004: 3), but the idea of 'benchmarking' spread to higher education from business in the early 1990s. The 1997 Report of the National Committee of Inquiry into Higher Education suggested that the newly established independent Quality Assurance Agency, executing the Teaching Quality Assessments since 1992, should provide 'benchmark information on standards' (Lund, 1998: 66). The value of these benchmarks would be to

> ensure that no matter where in Britain students choose to study for a geography degree, they would undergo broadly equivalent learning experiences, while the degree classifications they achieve as they graduate would be firmly based on some shared and public statement of standards.
> (Pepper et al., 2001: 24)

These benchmarks would not mean a 'national curriculum', although that was what was feared by some (e.g. Johnston, 1997), but a national system based on the principle of peer review to ensure broadly consistent quality and standards.

This administrative shift towards a formalisation of the quality assurance procedures in higher education, connected to increasing numbers of HEIs and students during the 1990s, also has implications for the position of the dissertation within the undergraduate curriculum. The dissertation is recognised as:

> particularly important to the issue of degree standards and academic quality, for it is central to (British) conceptions of quality in higher

education and generally contributed significantly to final degree classification.

(Pepper et al., 2001: 28)

Although almost all undergraduate geography students in the UK undertake a dissertation, there is some variation in terms of size, weighting, and timing of the work (Harrison and Whalley, 2008: 402). Looking at the most recent (at the time of writing) Geography Benchmark Statement, formulated by the QAA, it does presume 'something like a dissertation' as part of the degree:

> Within honours degree courses in Geography, it is anticipated that some form of independent research work is supported throughout the degree. Students experience the entire research process, from framing enquiry to communicating findings. Independent research is often communicated in the form of a dissertation presented in the later stages of the course. Other formats could include research posters, journal articles, vlogs or final reports and a choice of assessment format may be appropriate. Independent research may involve field-based data collection, or other forms of primary or secondary research, civic engagement or work placements.

(QAA, 2022: 16–17)

A dissertation as part of the undergraduate curriculum is thus not required by the benchmark statements and has never been required, but is nonetheless strongly approved and encouraged. For most British institutions, and the same goes for University of Glasgow, the dissertation, or 'individual research project' or 'final year essay', a fundamental part of the curriculum for decades, has acquired a further stamp of official-national sanction and, to an extent, direction.

The history and future of dissertations

Undergraduate dissertations are often seen as the key 'format' of undergraduate research. There is a difference between undergraduate research in the UK and in the USA. In the latter, there is a stronger tradition of undergraduate research by a selected group of students – not all students – working on student-initiated and faculty-supported research, whereas in the UK, the final-year dissertation is seen as the place and time for undergraduate research (Healey and Jenkins, 2009). The UK dissertation is hence an 'equal' experience for all students, meaning that it is expected from all enrolled students to produce a dissertation, and also that all these students are (or should be) supported in producing the dissertation. Work by educationalists on research undertaken by undergraduate students taking 'first degrees' focuses largely on the different dimensions that this research can take: it can be done by all students or just some, it can take the form of individual or collaborative projects, it can be process-centred or product-centred, and it can start in year one of the undergraduate degree programme or in the final year (Beckham and Hensel, 2009). The 'elite model' of just a 'lucky few' doing undergraduate research, a US tradition, is opposed by the traditional 'mainstream model' that is common in the UK. However, diving into what has been written *about* the phenomenon of the undergraduate dissertation, it soon becomes evident that more has been written about its expected or desired *future* than on its history. Research on the historical role of the undergraduate dissertation is in fact remarkably sparse. Some educationalists reflect very briefly on how the dissertation has been seen as a 'cultural expectation' (Healey and Jenkins, 2009: 19) for decades, and some succinctly address the distinctive historical role of the dissertation in curricula:

> The dissertation was regarded as the component of undergraduate studies that offered students the opportunity to demonstrate their "honour worthiness" 20 years ago. In courses in which much of the assessment was by examination, the dissertation was a relatively unique opportunity for independent learning and knowledge acquisition. In addition, the dissertation was designed to prepare students for postgraduate study.
> (Rowley and Slack, 2004: 176)

Such reflections, however, are very general and mainly address the presumed status of such independent study, and not about shifts in the exact forms, roles, and scope that the undergraduate dissertation has undergone through time. There are other ways to shine a light on such developments: first of all, of course, by looking at actual undergraduate dissertations from the past, which is one of the key contributions of this book. However, before turning to the undergraduate geography dissertations, it is insightful to look at rare instances of educationalists from the past discussing the (for these authors, *contemporary*) roles of the undergraduate dissertation or, as its senior cousin, the doctoral thesis: such publications then become historical sources, even if they were not written as such. Watson (1983), for example, researched dissertations as learning or teaching tools in the context of undergraduate business degrees in the UK. He describes thesis writing as regarded as indispensable, with courses seen as 'somehow incomplete if they do not include a thesis or dissertation' (Watson, 1983: 182). More recent educationalists describe the pressure on the 'classic' final-year undergraduate dissertation:

> For the last half century or more the final year undergraduate dissertation, typically an 8–10,000 word independent project, has been seen as the gold standard for British higher education. However, it is coming under pressure for reform as student participation rates have increased, the number studying professional disciplines has grown, and staff-student ratios have deteriorated. Some courses have abandoned the dissertation altogether, but there is a danger of throwing the baby out with the bath

water ... The dissertation has a long life yet. However, if it is to remain strong and vibrant and continue to provide a transformational experience for most students then it needs to evolve and become more flexible. We need to recognise that not all students want the same things from their degree programmes and that a choice of alternative or additional formats, experiences and outputs is desirable.

(Healey, 2011)

Again, this quotation – from an educationalist particularly concerned with geography and the wider environmental sciences – addresses the familiarity of the undergraduate dissertation for anyone involved in British academia, but no insights on the emergence of this 'rite of passage' is given.

Other than the undergraduate dissertation, the history of the PhD thesis has received more attention from historians goes far back, but the modern-day PhD in Britain finds it origin in 1917. Before then, British students went to German universities where they would be rewarded with a PhD, but for economic as well as political reasons this was not seen as a desirable situation anymore (Simpson, 1983 and Bogle, 2017). The first research doctorate programme in the UK was Oxford University's DPhil degree, and from the start a written 'dissertation' – later typically called a 'thesis' rather than a 'dissertation', although some interchangeability of the terms has persisted, even to the present – was a standard aspect of examination (Jackson and Tinkler, 2001). Another source, specifically on the 'growth' of geography as an academic discipline (Stoddart, 1967), addressed this growth by tracking, among other indicators, the number of 'theses accepted for higher degrees'. He experienced difficulties in finding data on the number of persons with academic qualifications, but solved this problem by using long-term data from the so-called 'Index to Theses Accepted for Higher Degrees in the Universities of Great Britain and Ireland', managing thereby to demonstrate exponential growth of geography in the UK from the late 1950s onwards (Stoddart, 1967: 6). Stoddart's research demonstrates how increasing numbers of higher or doctoral degrees might stand as a proxy for the growth of an academic discipline, as well as implying that dissertations and theses were increasingly viewed as standard elements of disciplinary curricula, training, and preparation. Of course, only higher degrees are mentioned by Stoddart, not specifically the BSc or MA degrees like the ones geography students at the University of Glasgow have obtained (as discussed later, geography students received a BSc or MA degree dependent on whether they were enrolled as science or arts students), but the contributions of both Stoddart and Watson suggest that dissertations and theses were standard, and to some extent unquestioned, by the second half of the twentieth century. However, the exact origin of specifically the *undergraduate* dissertation is never discussed.

The scarcity of literature and research on the history of the undergraduate dissertation itself, and on changes in the purposes, characteristics, and disciplinary differences of and associated with the dissertation over time, strikes as remarkable. The presence, though still limited, of research on the history of the PhD thesis might indicate that, rather more than the undergraduate degree, the PhD degree is seen as an 'entrance' to the academic community. Moreover, PhD theses are, of course, often the source of published books, chapters, and articles, as well as being quite commonly cited, and sometimes even discussed at some length, in the academic literature: as such, they are already there as part of the acknowledged textual record on which most histories of academic disciplines are based. The same is palpably not the case for undergraduate dissertations, even though undergraduate students have held a central role – in many respects an even more central role than PhD students – in the academic community.

Whereas Master's and Doctoral theses as well as their marking procedures are widely researched, academic attention to the undergraduate dissertation is thus sparse (Todd et al., 2004). Some researchers, however, have tried to formulate a key definition or description of what this particular educational 'format' exactly entails. One of these attempts to describe the undergraduate dissertation distinguishes four key characteristics: students define the focus of the work, the work is carried out on an individual basis (with support), students have prolonged in-depth engagement with the work, and research includes the analysis of primary and secondary data (Ashwin et al., 2017: 514). These four characteristics mention some important aspects of the process and scope of dissertation research, focusing on the responsibility and, at the same time, the freedom of students within this process. The possibility to choose a subject, to decide on methods, and to go to the field alone are fundamentally different from sitting an exam, and can thus be 'a liberating experience, allowing them [students] to display creativity and imagination' (Gatrell, 1991: 17). The specific character of the dissertation distinguishes it from other aspects of the undergraduate curriculum, a distinction not only experienced by students themselves but also by members of staff. The dissertations can be used to distinguish between degrees (e.g. honours versus ordinary degree) but are also drawn upon 'when it comes to adjudication of those falling on the borderline of degree categories' (Webster et al., 2000: 72). Although the notability of the dissertation within the undergraduate curriculum has remained similar over the decades, there have been some small, gradual changes. Rowley and Slack (2004) argue that before the 1980s, examinations were the dominant mode of assessment in most undergraduate courses, whereas towards the end of the twentieth century, coursework started more explicitly to assess knowledge as well as skills (Rowley and Slack, 2004: 177). The emphasis on independence and individual responsibility for student learning became part of undergraduate courses.

Dissertations as learning tool or assessment tool

Turning now to the local context at the University of Glasgow, which is the case study explored in this book, definitions of what the dissertation actually entailed over time can be found in the Syllabi of Examinations:

In the summer vacation students will undertake an independent regional study of an approved area. Approval of the area should be obtained by the end of the Candlemas term. The study, in the form of an essay not exceeding 6,000 words, is to be presented on the last day of the Martinmas term following.

(University Calendar, 1954–1955, SEN10/97: 261)

This citation presents some answers with regard to the requirements and rules concerning the study: this description mentions that it is an *individual* study, though the area which it is studying should be approved. The deadline and scope of the 'essay' are also mentioned. This description of the 'independent study' is more or less the same until 1970, when it is called 'a dissertation' for the first time. The Systematic Geography degree, however, speaks of 'a dissertation' from the early 1960s.

The question as to whether dissertation-writing students knew what was expected of them evokes follow-up questions about what overall is the *purpose* of the undergraduate dissertation. Are students 'sent' to the field without a lot of guidance, as if it should be self-evident based on everything that they have been taught, earlier in the undergraduate curriculum? From the Glasgow cohorts of the early 1950s to the cohorts of the 2010s, every one of these students has been on multiple collective fieldtrips during their undergraduate studies. From the early 2000s, the third-year geography field class was explicitly conceived as a training ground for the dissertation research. Is the dissertation the ultimate test as to whether students are 'worthy' geographers or is the project an opportunity to guide students through their first 'big' project?

Most will agree that it is not either-or, but considerations about whether the dissertation is an assessment tool or a learning tool strongly influence the relationship between student and supervisor, as well as the actual outcome of the research project. Frequently recurring discussions about the purpose of – and hence what preparation is needed for – the dissertation will doubtlessly be familiar to staff members of many geography departments: is the undergraduate dissertation primarily an *assessment* tool or a *learning* experience? The time span of the dissertation archive presents the possibility to analyse this 'assessment versus learning' question over time. Some interviews with current and former staff members gave more clues about how both students and supervisors saw the undergraduate dissertation. One alumna, who graduated from Glasgow in the early 1970s, reflects:

My memory is you got to hand it in the first day back, so we really did it over the summer. From May through to October, it would have been in those days. I am not conscious of any training for it, I am not conscious of having anyone to help me do it. Someone must have told us how to do it, but it could have been a handout for all I know.

(A. Dunlop, interview, 2019)

This is also recognisable in the sparse feedback comments saved from the 1960s and 1970s. Students are called to account on fundamental mistakes in the research design or research question, whereas for contemporary dissertation-writing students – hopefully – such mistakes will be overcome in an earlier stage of the research project. Yet again, it is not black and white, and a dissertation seen as an assessment can still be praised for its learning effects. In this 1966 marker's report, the marker praises the learning experiences of the mistakes made in the research:

> you had a lot of computing to do, and this has been done well – even more important however is the experiences gained by making all those mistakes.
>
> (Graham, Marker's Report, 1966)

The nature and differing conceptions of what a dissertation is or should be are, as shortly described here, also strongly related to one specific position: the undergraduate student's supervisor.

The role of the supervisor

The relationship between a supervisor and student is, to some extent, 'open-ended' (Derounian, 2011: 92). Although some of the key aspects of supervision are described in explicit documents about the dissertation, it is up to the supervisor and student to arrange a feasible working relationship – yet, increasingly with guidance – and all of this, in a relatively short period of time. MacKeogh (2006) distinguishes multiple roles taken by undergraduate dissertation supervisors:

> Subject experts; gatekeepers of academic standards; resource person and advisor on the research literature, research methodologies; 'midwife' of the dissertation; director, project manager, shaper; scaffolder and supporter; editor; promoter of student self-efficacy.
>
> (MacKeogh, 2006: 20)

The variety of these roles demonstrates the complexity of the question to what extent supervisors influence the final product of the research process, the dissertation itself. The supervisory relationship is shaped by balancing autonomy and support (Del Río et al., 2018). It is difficult to discover more about student–supervisor relationships based on individual dissertations themselves, yet Glasgow's dissertation collection as a whole reveals some developments over time of the role of the supervisor, or as they were often called before the 1970s, advisors.[1] First, the mentioning of the supervisor in the dissertations gradually became more frequent between 1950 and 2010. Whereas in the regional dissertations from the 1950s and 1960s, almost never was a supervisor mentioned (in line with the idea that the regional dissertation was a straightforward exercise), some of the systematic dissertations did mention a single academic member of staff, presumably the student's advisor. It was often 'just' by name, in a list of other parties and institutions that were thanked. From the 1990s on, the formalisation and indeed parallel intensification of the role of supervisor becomes evident in the dissertation archive. For instance, 'formal' dissertation record cards were used for a while. The much greater presence of the supervisor in the acknowledgements is also a window on increasing formal and informal forms of involvement:

> [Thanks] for all the chocolate, help and advice, without which this dissertation would not have been possible. I am sorry for all the times I e-mailed you on your day off but equally very grateful for all the times you replied. Hopefully I will have done you proud.
>
> (Gray, 2010: n.p.)

> I am unsure how many students are as lucky as I was to have such a dedicated and insightful tutor, who will allow them to work independently and self-sufficiently. For all your help, I am very, very grateful.
>
> (Davidson, 2014: 3)[2]

Acknowledgements often demonstrate a combination of intellectual support and counselling. Besides the interpersonal and advisory relationship between supervisor and student, shaped to some extent by the rules and guidelines of dissertation supervision and to a greater extent by the individual supervisor and student's preferences, the *intellectual* relationship plays a significant role as well. Students' dissertation projects were sometimes directly inspired by specific option courses, given by their supervisors. Some students mention this explicitly in the acknowledgements. Many more of these intellectual influences of particular supervisors (or generally, academic members of staff) become perceptible when studying the disciplinary and subdisciplinary trends and shifts in the undergraduate dissertations over time.

The broader departmental and intellectual network

The relationship between students and supervisors is not the only one that plays a role in both an intellectual and counselling way. The networks within the department, the wider university, and the broader academic community are also very much present in the undergraduate dissertations. Whereas the curriculum and all courses in the undergraduate studies have an effect on what students learn, some students explicitly mention one course or lecturer who has influenced their intellectual interests and skills. The frequency of 'non-supervising' academic staff members mentioned in acknowledgements demonstrates how the dissertation perhaps is the 'grand finale' of the undergraduate studies, but is simultaneously 'just' one component of a larger and much more extensive learning process, in which many different people in the department play a role. Supporting staff, such as lab and field technicians, are also mentioned by many students undertaking physical geography dissertations, especially the ones requiring an extensive period of fieldwork preparation or laboratory analysis within the department:

> [thanks to] Kenny Roberts for his assistance whilst working in the geography laboratory, especially for the much needed cups of tea!
> (MacKay, 2010: 3)

There are other departments of the University of Glasgow that were mentioned by some students: Zoology, Engineering, and Environmental Chemistry are mentioned the most. The latter provided research facilities that Geography did not have 'in-house'. Besides other departments from the university, some students also worked with individual staff members or used research facilities at other universities, both nationally and internationally. Lastly, peers play a fundamental role in the experience of undergraduate dissertation writing. As mentioned earlier, some students worked together in the field; for instance, in one of the field expeditions. Yet, also 'just' being in the writing-up stage and sharing the same deadlines and similar struggles turned out to be memorable experiences for students:

> [Thanks to] All the guys and gals in my class who helped me in my hour of need (4 pm–5 pm, Monday September 1st) on the Macs, on maps, stats and on inspiration which was badly needed:- Elaine, Helen, Kieth [sic], Laurence, Lesley, Marion, Paul, Pauline and Roy.
> (Martin, 1986: iv)

This is perhaps one of the most recognisable experiences for all former undergraduate dissertation writers (all current academics, and many, many more people now with a career outside of academia): the hours, days, and weeks of writing-up, having coffee breaks together, and offering each other small forms of mental and intellectual support. The university library has been a central place to many dissertation writers over all decades, both as a place to find source material and as a place to meet others going through the same process.

Lastly, there is an important group of people who students barely acknowledge at all: all the former thinkers on whose work they are building. The undergraduate dissertation, then, is indeed very similar to other academic research projects, with the historical and contemporary ideas of other academics as starting point, with many peers or colleagues working on similar yet different projects, and the daily struggles of writing, distraction, and motivation. The distinction perhaps is found in two ways: the inexperience of doing a project like this individually, and the educational context of needing to do this project within certain requirements in a certain period of time to graduate.

Increasing explicit governance of the dissertation process

In the mid-1990s, the recording of the supervision process became more formalised. It is difficult to pinpoint one specific reason for these changes in supervision, as it was a combination of growing student numbers (leading to a bigger workload in terms of supervision), professionalisation and quality assurance in higher education, and individual staff changes. Reflecting on these changes, two former academic staff members mention some aspects of this process within the department:

> The naughty difficulty of how much support would you give to students…some students were getting a lot more support than others. There was no malice in that, it was just that some supervisors thought, this is the way to do it. We tried to codify it more, there were several things: some dissertation topics led themselves more easily to supervision, because they were intrinsically structured. Some students didn't bother very much about supervision. Gradually we became more and more involved.
>
> (G. Dickinson, interview, 2019)

> When we were in the phase that we saw the dissertation very much as an assessment vehicle, one of the things we noticed was that you could flick through them and you could see quickly quite a lot of mistakes. We did, ten years ago, introduce a process that we still use where students can have part of all of their dissertation skim-read by their supervisor before they hand it in, about 10 days, 2 weeks before, just to get them quick feedback on the obvious mistakes: reference list incomplete, no captions for your figures, spelling mistakes, whatever. … I think what happens with the best students they almost don't need it, because having that meeting makes them prepare for it, which means they would come along with something that is actually pretty good, and you give them almost trivial feedback. But if you wouldn't have the meeting, they wouldn't have done that. It is forcing them to be prepared for that meeting that is actually … Quite a few of the changes we made over the years have been designed to do that, putting milestones in, making the students do the reporting.
>
> (T. Hoey, interview, 2019)

Some cohorts of the late 1990s in the archive include a dissertation record card. This double-sided card addresses the dissertation process, from getting information about dissertation writing and subject choice to handing in the dissertation. These record cards provide a lot of information on both the formalities of dissertation writing and the differences in the actual execution of all dissertation-related tasks. It demonstrates the lengthiness of the dissertation process, from an introductory lecture about the dissertation at the start of the second term of the junior honours year (year 3) to handing in the completed dissertation to the departmental office on day 1 of next year's second term.

The record card also explains how an initial dissertation idea, often based on one of the honours options, had to be submitted to the dissertation administrator, after which, two weeks later, the student would get the name of the appointed supervisor. The student and supervisor then usually met twice before summer: in the last meeting, the dissertation plan made by the student should be formally approved by the supervisor. The student usually undertook the fieldwork for the dissertation in the summer between third and fourth year, an expectation and practice consistent for the cohort of 1954 up to the cohort of 2020 and beyond. The record card itself was not a formal aspect of assessment, but as the card states: 'A record which shows that the student completed all compulsory meetings before the deadlines, will count in his/her favour'. It seemed to have played a role in cases in which the marks of two dissertation markers on average was in the middle of two grades. From approximately the mid-2000s, a formal dissertation proposal is an assessment that gets marked as part of a student's third year (junior honours). The dissertation record card as well as the appearance of something like a 'Dissertation Handbook', including grade-related criteria and an overview of rules and guidelines concerning the undergraduate dissertation in the mid-2000s demonstrates the governance becoming increasingly explicit over time. Unfortunately, earlier versions of hand-outs of dissertation requirements are not saved and archived. The consistency in, for instance, the scope and format of dissertations within cohorts suggests some rules and guidelines. However, it is not possible to discover how implicit or explicit these forms of governance were.

The dissertation and the curriculum

The undergraduate dissertation is of course the capstone of years of study, and it is important to connect the dissertation and the knowledge produced in such works to the undergraduate curriculum as a whole. By looking at the shifts in the curriculum, it becomes obvious how students' different experiences and final 'knowledge products' are inextricably linked to the years before. Looking at one cohort from the mid-1950s, and the prescribed literature, it gives some insights into the content of their undergraduate education, see Table 2.1.

The regional focus of the degree is evident, with many sources on specific regions in the world such as The British Isles, Latin America, the U.S.S.R, and Western Asia. The other literature is about subdisciplinary fields such as geomorphology, cartography, economic geography, and historical geography. Third-year students were taught the history of geographical knowledge, as the mentioning of Dickinson and Howarth's *The Making of Geography* (1933), itself already terribly dated by then, indicates. This work provides an overview of the history of geographic thought, from the ancient Greeks, via the colonial era up to the development of 'the regional concept', and overall, the curriculum

Table 2.1 Prescribed books for academic year 1954–1955 (further readings recommended during the course) (University Calendar, 1954–1955, SEN10/97)

Ordinary Class – Year 1
Bartholomew	The Advanced Atlas of Modern Geography
Lake	Physical Geography
Wilmore	Groundwork of Modern Geography
Hardy	The Geography of Plants
Shackleton	Europe – A Regional Geography
Demangeon	The British Isles

Higher Class I – Year 2
Jones and Bryan	North America
Baulig	Amérique Septentrionale
Stamp	Asia
Sion	Asies des Moussons
De Martonne	Traité de Géographie Physique
Von Engeln	Geomorphology
Raisz	General Cartography
Miller	Climatology
Kendrew	The Climates of the Continents

Higher Class II – Year 3
Gregory and Shave	The U.S.S.R.
Blanchard	Asie Occidentale
Preston James	Latin America
Fitzgerald	Africa
Wood	Australia – its Resources and Development
Forde	Habitat, Economy and Society
East	An Historical Geography of Europe
Dickinson and Howarth	The Making of Geography

Higher Class III – Year 4
Vidal de la Blache and Gallois	Géographie Universelle (European Volumes)
Ogilvie	Great Britain, Essays in Regional Geography
Smith	An Economic Geography of Great Britain
Sorre	Les Fondements de la Géographie Humaine
Zimmermann	World Resources and Industries

strongly emphasises *regional* geography. What this list of literature reveals, besides the regional focus, is the inclusion of literature published in languages other than English. All geography students should take at the end of second year:

> At the end of 2nd year, intending honours students will be required to pass a test in translating, with the aid of a dictionary, a passage from a geographical text in either French or German
> (University Calendar, 1954–1955, SEN10/97: 261)

From the late 1960s, this language test was not confined to either French or German, but could also be taken in Spanish, Portuguese, or Russian (SEN10/113: 247). A few years later, this language test was no longer mentioned in the University Calendars and was thus probably taken out as a requirement before passing to Honours.

Whereas in the 1950s, the undergraduate geography degree was called 'Geography', without any distinctions between regional or systematic geography, from the early 1960s, the degree called geography could be combined (as a double course, predecessor of joint honours) with systematic geography.[3] This is also when the first students with two dissertations in the archival collection appear; for instance, a regional study entitled *The Parishes of Falkland, Leslie and Kinglassie* (Balfour, 1962a) combined with a 'systematic' urban geography dissertation entitled *A study of functional change in Park District* (Balfour, 1962b), was undertaken by the same student. During the 1960s and 1970s, the curriculum changed from being very 'fixed', with little space to choose courses within a certain degree, to a programme with an increasing number of choices and options. For instance, systematic geography students could choose classes on biogeography, urban geography, economic geography, historical geography, and climatology, besides the mandatory course on air photo interpretation. The systematic dissertation, then, had to be connected to one of the courses chosen. From 1970, Honours Students were taught the course Development of Geographical Thought, and in 1974, systematic geography was extended with the courses applied human geography and social and political geography. In the late 1970s, the distinction between geography and systematic geography disappeared, and the only undergraduate geography degree was called 'Geography' once more. Within this course, students could still opt for a more regional or systematic focus by their choices of option courses. In 1980, a limit on regional options chosen was formulated.

As mentioned, from the early 1960s, the geography degree could be combined, as a double course, with systematic geography. Whereas in the 1950s there were solely single courses, the development towards double courses, later called joint degrees, was one to stay. In the early 1960s, geography could be combined with systematic geography, political economy, economic history, archaeology, and computing science. In later years, new potential combined degrees with geography emerged (and sometimes disappeared again after a few years): during the 1980s, for instance, combinations with Arabic, philosophy, and Scottish literature became possible. These potential joint degrees do not reflect the curriculum of the undergraduate geography degree, but they do indicate changes in the university: with many new degrees and new combinations of degrees, studying had become a more 'tailor-made' experience, whereas geography students in the 1950s, after having chosen their degree programme, were part of a relatively static four-year programme.

Dissertations and authorship

Studying an archive as one collection creates an imagined community that transcends time (Daston, 2012). Because these dissertations all ended up in one collection in one cupboard, every individual archival source has its own context and history. Although the reason why every dissertation in the archive ended up in the archive is because of its 'belonging' to the selection of the archive (in this case, being an undergraduate geography dissertation, produced at the University of Glasgow), the individual dissertations should still be acknowledged as individual narratives, with individual contexts, connected to specific travels, emotions, and personal relations. Although there are some fundamental differences between undergraduate dissertations and books – for instance, the way they are circulated – they share an *active* constitution: 'the negotiated and contested outcome of the interplay of material and social processes' (Keighren, 2013: 745). The fundamental differences are related to the power dynamics at play between, on the one hand, often relatively young students, and, on the other hand, older staff with a certain academic status, that have the task to supervise or grade the dissertation produced by the student. The rules and guidelines that are formulated by 'the university' as an institution also have an impact on the way dissertations are constituted. In terms of circulation, dissertations – unfortunately! – usually only have a very small audience, but this audience is often closely involved with the author: supervisor(s) and marker(s), peers, friends, and family constitute the vast majority of the readers.

There is one remark that should be made in the discussion of authorship. The dynamics between both seeing every individual student who wrote her or his dissertation as an independent researcher, as well as constantly being aware of the social and institutional context in which the dissertation came about, raises the question: whose words, ideas, definitions, and thoughts do we read when reading a dissertation? Undergraduate students were influenced by, and reciprocally influenced, classmates, supervisors, markers, the offered courses in the undergraduate curriculum, and university rules and guidelines. Such influences can of course also be seen in the research produced by 'established' academic geographers when speaking to direct colleagues, students, at conferences, and getting reviewed by journal editors, but are even stronger in the case of undergraduate students. It might be hard to contradict things one has learnt from one's direct supervisor or to question an idea broadly carried out in the undergraduate curriculum, especially when one is (often) both significantly younger than the ones with the 'final judgement' and dependent on graduating in financial (fees and loans) and social (expectations of peers, parents, and oneself) ways. It is a very strong playing field of power relations that cannot be ignored:

> ... when we approach a printed text from the past, we need to consider not just who wrote it – the biography of the author, their aims and ambitions – but also when and where the text was written and what that tells us of their notion of being an author and what could and could not be done/expressed by an author.
>
> (Mayhew, 2007: 26)

Indeed, given the aim to investigate the relational context of which 'small' knowledge *producers* were a part, the role of the university department overall, with its staff members, curriculum, and rules and regulations, as well as the big 'outside world', and one particular category of 'relations' that are considered especially important in this regard is the role of peers, meaning fellow students who share the same educational context and perhaps the same study rhythms and routines and social lives. By studying the dissertations of a complete yearly cohort of students, it is hence possible to discern the relational connections within the cohort, rendering a more sociological perspective on this production of geographical knowledge more achievable:

> The most important of these [geographical elements of cohortness] is the sense of thrown-togetherness – or, as we have specified, *synchronicity* – that emerges in the ways in which individuals are corralled into institutional and other collective spaces. We have been at pains to emphasise how the everyday, emotional, material experience, and *performance* of that synchronicity can at one be accidental and deliberate, and can at once lead to both senses of belonging and exclusion.
>
> (Brown and Kraftl, 2019: 14)

Conclusion

The educational context of the undergraduate dissertation as part of a wider curriculum distinguishes such a research project from other academic research. It is not only the awareness of students and supervisors that the dissertation should be 'good enough' to pass (or to get a certain aspired grade) but also the relationship between what the students had learnt and which fields the students had come to know well during their coursework that in many cases directly influences the dissertation. Specific courses, individual staff members, and existing facilities within the department or university might influence the choices made by students. These aspects mean that, whereas many undergraduate geography students in the UK might research similar phenomena in similar 'fields', there will always be substantive differences in when certain disciplinary trends become visible in undergraduate dissertations or which areas are more regularly studied. The 'where' of dissertation research is multi-layered and diverse. The more than 2,600 small knowledge productions of past geography students at the University of Glasgow all include quite specific geographies of how the research for the project is undertaken and experienced, yet taken together, this collection has revealed numerous key geographical, social, and educational constants and changes over time.

The specific education context distinguishes the undergraduate dissertations to some extent from other geographical knowledge productions and raises questions about *who* exactly produced the content of the dissertations. It is important to be aware that the dissertations – the first independent contributions to the field of geography – by 'geographers-in-the-making' are not only worth studying because of their sometimes innovative and well-researched foci of inquiry and methods but also because they offer a glimpse into how 'established' geographers impact future generations of geographers and how they 'made an entrance' into the discipline themselves. The role of supervisors and other staff members; the networks of peers, friends, and family; the financial and practical access to research further from home; and the rules, regulations, and traditions of British higher education are central to the productions of knowledge and 'rites of passage' that are undergraduate dissertations.

Intermezzo 2: Outdoor Recreation

There are numerous dissertations undertaken about outdoor recreation. These take different perspectives: sometimes emphasising the capacity for and impacts of outdoor recreation in a certain area, other times the relationship between recreation and conservation, and still on other occasions people's recreation preferences. The influence of individual academic staff members is perceptible in relation to this topic. Dickinson, a lecturer in geography at Glasgow from the 1970s to the 2000s, supervised many dissertations about Loch Lomond and outdoor recreation. The emphasis in many of these projects was on the difficulties of balancing the respective environmental 'needs' of recreation and conservation. One of the projects supervised by Dickinson was Frances Jarvie's dissertation *The capability for outdoor recreation of south-west Stirlingshire* (1974). In this study, she analysed the 'capability' of environments to support different kinds of recreation such as caravanning or walking. Such an analysis was then mapped, as seen in Figure 2.1: the overall score in one square being the sum of a score from 1 to 3 for slope, texture, and drainage, leading to an overall value between 3 and 9 quantifying the extent to which each small area, or square, could support an activity such as caravanning.

These maps, organised per 'recreation activity', then formed the basis for the conclusion of the dissertation, combined with a survey asking participants to rank 14 photographs in order of preference with respect to scenery.

Another example concerning the relationship between recreation and the physical landscape was Rhona Thomson's dissertation entitled *Recreational erosion of the vegetated dune environment, Alagadi Beach, northern Cyprus* (1998). The focus of this dissertation also connects recreation and the physical landscape, but is somewhat the opposite in framing to Jarvie's dissertation: Thomson's starting point was the 'carrying capacity' of the landscape, whereas Jarvie took the needs of recreation activities as her starting point:

> The demand for public outdoor recreation, particularly in coastal areas has increased significantly in recent decades. This new trend has, however, been accompanied by negative consequences. Increase in recreational

32 *The history of geography and educational practices*

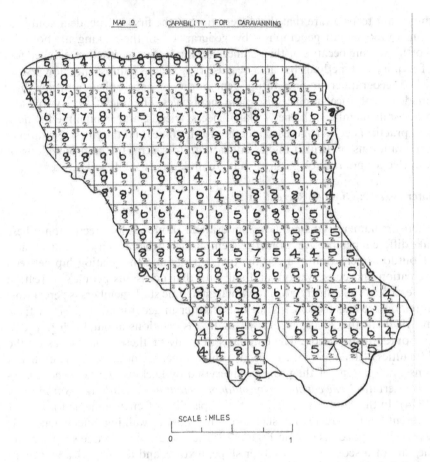

Figure 2.1 Map of 'caravanning' capability, based on slope, texture, and drainage (Jarvie, 1974).

activity results in an increase in the human presence in these areas which in many cases can cause severe degradation of the natural environment.

(Thomson, 1998: 1)

Recreation and conservation are hence often studied hand-in-hand, but the focus can differ greatly. There is no specific trend or change over time that is readily identifiable, and studies here are perhaps mostly formed by the student's own interests.

Notes
1 This term has periodically reappeared, probably to suggest a more 'hands-off' relationship: *advising* the student as explicitly different from *supervising* the student.
2 There is an interesting linguistic difference noticeable from the late 2000s, when students address their supervisors directly in the acknowledgements, as if 'speaking to them'.

3 'Systematic' geography is equated with a sub-disciplinary focus and was often seen as a complement to regional geography. In the latter, all systematic geography would supposedly be put together with reference to specific 'areas' or 'regions'.

Bibliography

Ashwin, P., Abbas, A., McLean, M., 2017. 'How does completing a dissertation transform undergraduate students' understandings of disciplinary knowledge?' *Assessment & Evaluation in Higher Education*, 42(4), 517–530.
Balfour, M.E., 1962a. *The Parishes of Falkland, Leslie and Kinglassie, Fife*. Undergraduate Dissertation, University of Glasgow.
Balfour, M.E., 1962b. *A study of functional change in Park District*. Undergraduate Dissertation, University of Glasgow.
Beckham, M., Hensel, N., 2009. 'Making explicit the implicit: Defining undergraduate research'. *Council for Undergraduate Research Quarterly*, 29(4), 40–44.
Bogle, D., 2017. *100 years of the PhD*. https://www.vitae.ac.uk/news/vitae-blog/100-years-of-the-phd-by-prof-david-bogle
Brown, G., Kraftl, P., 2019. 'Theorising cohortness: (Mis)Fitting into student geographies'. *Transactions of the Institute of British Geographers*, 44(3), 616–632.
Brown, R., 2004. *Quality Assurance in Higher Education: The UK Experience Since 1992*. London: Routledge.
Daston, L., 2012. The Sciences of the Archive. *Osiris*, 27, 156–187.
Davidson, D., 2014. *The winds of change - A changing landscape within the context of renewable energy*. Undergraduate Dissertation, University of Glasgow.
Del Río, M.L., Díaz-Vázquez, R., Maside Sanfiz, J.M., 2018. 'Satisfaction with the supervision of undergraduate dissertations'. *Active Learning in Higher Education*, 19(2), 159–172.
Derounian, J., 2011. 'Shall we dance? The importance of staff-student relationships to the pursuit of undergraduate dissertations.' *Active Learning in Higher Education*, 12(2), 91–100.
Dickinson, R.E., Howarth, O.J.R., 1933. *The Making of Geography*. Oxford: Oxford University Press.
Ferguson, A., 2014. *A paranoid nation in film: The internalisation of external geopolitical threat*. Undergraduate Dissertation, University of Glasgow.
Gatrell, A.C., 1991. 'Teaching students to select topics for undergraduate dissertations in geography'. *Journal of Geography in Higher Education*, 15(1), 15–23.
Gray, S., 2010. *An investigation into the social and spatial worlds of the elderly in two contrasting areas of Glasgow*. Undergraduate Dissertation, University of Glasgow.
Hall, T., Toms, P., McGuinness, M., Parker, C., Roberts, N. 2015. 'The changing administrative place of Geography in UK higher education: Where's the geography department?'. *Area*, 47(1), 56–64.
Haraway, D.J., 1991. *Simians, Cyborgs and Women: The Reinvention of Nature*. London: Free Association.
Harrison, M.E., Whalley, W.B. 2008. 'Undertaking a dissertation from start to finish: The process and product'. *Journal of Geography in Higher Education*, 32(3), 401–418.
Healey, M., 2011, 28 June. 'Rethinking the undergraduate dissertation'. *The Guardian*. https://www.theguardian.com/higher-education-network/blog/2011/jun/28/flexible-dissertations-for-undergraduates
Healey, M., Jenkins, A., 2009. *Developing Undergraduate Research and Inquiry*. York: Higher Education Academy.

Jackson, C., Tinkler, P., 2001. 'Back to basics: A consideration of the purposes of the PhD viva'. *Assessment and Evaluation in Higher Education*, 26(4), 355–366.

Jarvie, F.A., 1974. *The capability for outdoor recreation of south-west Stirlingshire*. Undergraduate Dissertation, University of Glasgow.

Johnston, R.J., 1997. 'Graduateness' and a core curriculum for Geography?' *Journal of Geography in Higher Education*, 21(2), 245–251.

Keighren, I.M., 2013. 'Geographies of the book: Review and prospect: Geographies of the book'. *Geography Compass*, 7(11), 745–758.

Lorimer, H., Spedding, N., 2002. 'Excavating geography's hidden spaces'. *Area*, 34(3), 294–302.

Lund, H., 1998. 'Benchmarking in UK Higher Education'. In: UNESCO. *Benchmarking in Higher Education*. Paris: UNESCO, 66–92.

Mackay, H., 2010. *Climatic impact of the last glacial-interglacial transition on Loch Garten, Cairngorms, Scotland*. Undergraduate Dissertation, University of Glasgow.

MacKeogh, K., 2006. 'Supervising undergraduate research using online and peer supervision'. In: Huba, M. (ed). *International Virtual University Conference, Bratislava 14–15 December 2006*. Bratislava: Technical University Bratislava, 19–24.

Martin, J., 1986. *Unemployment and provision of leisure facilities in Glasgow*. Undergraduate Dissertation, University of Glasgow.

Mayhew, R.J., 2007. 'Denaturalising print, historicising text: historical geography and the history of the book'. In: Gagen, E., Lorimer, H., Vasudevan, A. (eds). *Practising the Archive: Reflections on Method and practice in Historical Geography*. London: Royal Geographical Society, 23–36.

Pepper, D., Webster, F., Jenkins, A., 2001. 'Benchmarking in Geography: some implications for assessing dissertations in the undergraduate curriculum'. *Journal of Geography in Higher Education*, 25(1), 23–35.

Philo, C., Lorimer, H., Hoey, T., 2009. 'Guest editorial'. *Scottish Geographical Journal*, 125(3–4), 221–226.

QAA, 2022. Subject Benchmark Statements: Geography. https://www.qaa.ac.uk/the-quality-code/subject-benchmark-statements/geography

Rowley, J., Slack, F., 2004. 'What is the future for undergraduate dissertations?' *Education + Training*, 46(4), 176–181.

Simpson, R., 1983. *How the PhD Came to Britain: A Century of Struggle for Postgraduate Education*. Guildford: Society for Research into Higher Education.

Stoddart, D.R., 1967. 'Growth and structure of geography'. *Transactions of the Institute of British Geographers*, 41, 1–19.

Thomson, J.L., 1998. *The socio-sypatial relationships of children in Livingston*. Undergraduate Dissertation, University of Glasgow.

Todd, M., Bannister, P., Clegg, S., 2004. 'Independent inquiry and the undergraduate dissertation: Perceptions and experiences of final-year social science students'. *Assessment & Evaluation in Higher Education*, 29(3), 335–355.

Watson, D.M., 1983. 'Dissertations as a learning and teaching tool: Undergraduate business studies in the UK'. *Improving College and University Teaching*, 31(4), 182–186.

Webster, F., Pepper, D., Jenkins, A., 2000. 'Assessing the undergraduate dissertation'. *Assessment & Evaluation in Higher Education*, 25(1), 71–80.

Withers, C.W.J., 2009. 'Place and the "Spatial Turn" in geography and in history'. *Journal of the History of Ideas*, 70(4), 637–658.

3 Spatial contexts of student knowledge production

The expanded geographical field

From the 1970s onwards, historians of science began to account for the spatial element of making science, an endeavour to which – by the 2000s – historical geographers started usefully to contribute to. Yet, the extent to which this perspective was then turned *back* upon the history of geography itself, as academic discipline or more diffuse subject, has actually been quite patchy. There have been some exceptions, initially ones mainly focused on one specific scale of inquiry, namely studying geographical traditions at different national scales (Livingstone, 1995), but it is only relatively recently that the geography *of* geography – or the historical geography *of* geography – has opened up to all manner of possibilities, including taking seriously a diversity of routes and sites in the manner just identified. There is a distinction sometimes made between places that geographers *research* and those that they research *in*, the former being incredibly broad, while the 'binary divide' (Cloke and Johnston, 2005) in geography and its associated practices – loosely, between human and physical geography – broadens the spatial range of the latter even more. Some geography 'happens' in the laboratory, an exclusively scientific domain. However, a lot of the research undertaken in geography takes place outside, in 'the field'. This distinction between 'closed-system' research and 'natural field locations' (whether these are actually more natural or more cultural) is an important one, because it distinguishes exclusively scientific spaces from other kinds of spaces: 'Unlike laboratories, natural sites can never be exclusively scientific domains. They are public spaces, and their borders aren't rigorously guarded' (Kuklick and Kohler, 1996: 4). The different foundations of natural sciences, on the one hand, and most of the social sciences and the humanities, on the other, are hence strongly connected to questions about status and credibility (Kohler, 2002). An exclusive scientific space makes research verifiable, replicable, and supposedly certain, whereas just another space out there 'in the field' is contingent, unstable, and susceptible to all kinds of internal and external influences. The field, however, offers possibilities concerning the involvement of the public as 'incidental' research assistants: not as hired field assistants, but just by residency (Vetter, 2011). Such assistants are essential to the production of scientific knowledge, but are usually hidden from the public view (McCook, 1996): we encounter a few of them in the undergraduate dissertations. The university department connects with other places, for instance, by researchers presenting their work in other universities or at academic conferences and by the diffusion of ideas by means of monographs, journal articles, and tweets. The balance between researching and teaching is a difficult one; yet, looking at the sheer number of people 'inhabiting' the departments, undergraduate students vastly outnumber staff members. All these students travel to places for their dissertation research (nearby or far away) and talk to family, friends, supervisors, peers, and people who they meet on their travels. In conceiving such a bottom-up narrative for inspecting the history of geography, the knowledge productions of all these students as well as their practices of becoming a geographer are central.

By looking at the small knowledge productions of many cohorts of geography students, in this case, of Glasgow University, it is possible to identify some general shifts and trends. For instance, issues of globalisation will be explored in relation to the affordability of travelling abroad as well as the internationalisation of higher education, perceptible both in a more diverse student population in the department and in more undergraduate students going abroad as part of exchange programmes such as ERASMUS, are all traceable in the dissertation archive. Connected to the whereabouts of study areas are also more conceptual discussions concerning the conception of what 'the field' actually comprises. These themes are extensively discussed by professional geographers. By taking the changing practices of dissertation-writing practices as a starting point, however, a different light can be shed on how exactly these theoretical and conceptual discussions about the field are expressed in academic geography education. 'The field' has held different appearances for different students. The practicalities of fieldwork and the role of collective, organised field expeditions played a significant role in the dissertation-writing experiences of many past students. The study of spatial practicalities addresses the geographical situatedness of small research projects.

The geographies of undergraduate dissertations

The undergraduate geography degree programme obviously has to do with 'the spatial', whether it is about spaces and places nearby or far away. For many students, this meant that travelling to their so-called 'study areas' to conduct their research 'in the field' was an integral part of their undergraduate experience. In a few cases, the student had not actually travelled to this location, since by some students the research was done 'from a distance', for instance, by doing a library study.[1] In most cases, physical encounters with study areas have been undertaken. The extensive dataset of over 60 years of geography student cohorts' independent works provides opportunities to recognise changes over time and discuss potential causes of shifts in such study areas. Table 3.1 demonstrates the numbers of dissertations about an area (or using an area as case study) within Scotland, the rest of the UK, and outside of the UK, respectively: it illustrates the overwhelming majority of dissertations about an area 'close to home' or, at least, close to university.

Table 3.1 Number of dissertations per study area category

Category	Number of dissertations
Scotland	1811
Rest of the UK	295
International	446
Other	62
Total	2614

When looking at the shifts in study areas over time, it is striking that dissertations with a study area in Scotland have always been the majority of the dissertations in any cohort. The number of dissertations about an area in the UK but outside of Scotland remains relatively constant but there is a shift in the ratio of Scottish versus international study areas over time. There is a notable dropping off of non-Scottish study areas from the mid-1970s to the early 1990s. It might be the case that with gradually expanding of education, more students of lesser financial means were studying at the university. Overall, the demographic of students at the University of Glasgow has predominantly been the West of Scotland (Thompson et al., 2009), and this is still, to lesser extent, the case.

There is a strong change towards more international study areas since the early 1990s. There are two apparent causes for this shift, one related to a broader development in higher education, the other more specific to the geography department at the University of Glasgow. First, in the early 1990s, internationalisation became a key issue in debates and policies concerning UK higher education. Internationalisation was 'expected to serve peace and mutual understanding, quality enhancement, a richer cultural life and personality development, the increase of academic quality, technological innovation, economic growth and societal well-being' (Teichler, 2009: 95). Student mobility was conceived as one of the key elements of internationalising higher education (ibid.). In 1987, the ERASMUS programme was established as an exchange programme between 11 European Union Member States, including the UK (European Union, 2012). Unfortunately, the number of geography students at Glasgow who took part in an ERASMUS exchange are not available, but presumably the emphasis on student mobility was present in the department as well. It is expected that, either by means of ERASMUS exchange and related financial support to go abroad or by the fact that going abroad for studies became common, geography students in Glasgow would have shown similar signs of international mobility. The second cause of this shift towards more international undergraduate dissertation study areas is easily identifiable: individual academic staff members in the department organised research trips to several different countries. For instance, from 1992 to 1995 there were annual expeditions to Iceland, organised by David Evans, with many undergraduate students joining for their dissertation research. Also perceptible in the graph are the research trips to Egypt and Tanzania, organised on several occasions by John Briggs and Jo Sharp, between 2002 and 2013.

Although Table 3.1 categorises the dissertations by their study area, making use of a threefold classification of Scotland, Rest of the UK and International, there is a fourth group of study areas yet not discussed. This category of 'Other' study areas entails approximately 2.5% of the total number of dissertations, but this collection reveals some interesting shifts. The category 'Other' consists of several kinds of dissertations: methodological dissertations, dissertations studying at the global scale, dissertations that are comparing case studies in Scotland with case studies in other countries, and dissertations about the virtual world. Although the number of methodological dissertations is relatively small, they do appear throughout the years in the dissertation archive; there is no specific trend perceivable. The dissertations emphasising the global scale or the virtual world are all from recent cohorts, appearing from the cohort of 2006 onwards. The emergence of dissertations on the virtual world are obviously connected to developments in technology, whereas the dissertations that address a question or research subject on a global scale are presumably connected to ideas, practices, and the clear academic importance and addressing of globalisation.

Study areas in Scotland

When addressing the more specific study areas of dissertations written about an area in Scotland, it is striking to note that many study areas have a direct connection to specific sub-disciplines, with many dissertations about Glasgow fitting into the categories of, for instance, urban geography, and many dissertations about the Cairngorms focusing on geomorphology. Some other study areas, however, demonstrate a much more dispersed distribution. A noteworthy development is the emergence of dissertations on Scotland on a national scale. Before the turn of the millennium, this was rather unique whereas since 2000 this has become more common. Since then, almost a quarter of all dissertations on an area within Scotland took a national perspective. The titles of dissertations about Scotland on a national scale or with a national perspective demonstrate some specific foci of inquiry: for instance, Scottish identity (both political and cultural), Scotland and sustainability, and Scottish cultural and environmental heritage. Although in some cases the scale of Scotland was a very practical one (for instance, when collaborating with an organisation such as Scottish National Heritage), in many cases the research on a national Scottish scale can be seen as exemplary for the rise in a Scottish cultural as well as political identity: approaching Scotland more as a nation-state than before, as also addressed in other academic publications (e.g. Leith and Soule, 2011). With political developments such as the foundation of a Scottish Parliament in 1999 (Mitchell, 2014), Scotland as a nation-state was more evidently a focus of inquiry for academics, including student-geographers.

Table 3.2 Number of dissertations per study area within the UK (but outside of Scotland)

Country/Crown dependency	Number of dissertations
Channel Islands	2
England	176
Isle of Man	3
Northern Ireland	40
UK National	61
Wales	13
TOTAL:	295

Study areas in the rest of the UK

The 295 dissertations about an area within the UK but outside of Scotland, can also be divided into multiple spatial subcategories. Table 3.2 demonstrates the broad coverage of areas examined by undergraduate geography students throughout the years. Similar to the overview of Scottish study areas, the broader national perspective is again a substantive category. These dissertations with a British scope take, in large majority, a human geography perspective: approximately 75% of these dissertations consist of cultural, economic, political, and social geography research. This proposition again might not be surprising: the nation-sate as a scale is feasible in research on politics as well as questions concerning identity, traditions, media, and migration.

Whereas one may have expected Northern Ireland to be an area that would be discussed through a political or cultural perspective in dissertations, this turned out to be incorrect. Although the Troubles were mentioned (as a contextual element) in some of the dissertations about Northern Ireland, only five of them had the Troubles, or the Northern Irish/Irish border conflicts, as their focus of inquiry. The connection between the Troubles and tourism was the main focus of the other two dissertations on the Troubles. Overall, the dissertations about an area in Northern Ireland had principally either a tourist or an agricultural emphasis. The spatial range of dissertations written about England is large. From Cornwall to Norfolk and from Hampshire to Northumberland, all counties are covered within the undergraduate dissertation archive. The spread is large as well. There are no outliers, except that the most northern counties of England are more often researched than the southern counties; the counties Cumbria, Northumberland, Tyne and Wear, Durham, Lancashire, and Yorkshire take up around 30% of the dissertations on England. There is no specific sub-disciplinary trend perceivable in the dissertations about the north of England except a small 'peak' of physical geography dissertations situated in Yorkshire. As demonstrated in these examples, this does not indicate a specific sub-disciplinary trend as the dissertations are on fluvial and glacial geography as well as coastal geography. The higher number of dissertations on the north of England thus seems to be based on its closer proximity to Scotland or the home areas of students instead of a specific sub-disciplinary focus.

International study areas

The broad range of study areas within Scotland and the rest of the UK is also perceptible in the diversity of countries, oceans, or political units (such as the European Union) in the rest of the world. There have been 71 different such units studied. Looking at the 'top 10' countries studied (see Table 3.3), there are some clear trends perceptible.

There are two outliers: Tanzania and Iceland. The high numbers of these two countries, together with Egypt, Switzerland, Norway, and New Zealand (also in the top 10), can be explained by the fieldtrips and expeditions organised by several particular staff members. All of the 69 dissertations on Tanzania, for instance, were written by students in the cohorts between 2008 and 2013 and have a human geography emphasis or emphasise a more hybrid form of geography (combining human and physical elements). Other countries that are more common within the dissertation archive, such as the USA, Ireland, Canada, and France, do not have a direct curricular or extracurricular cause for their higher presence in the collection, but the spatial proximity or the sharing of a native language are potential explanations. When international dissertation research became more common from the early 1990s, the diversity of countries visited increased as well. The majority of the students from the 1960s–1980s who travelled abroad went to France or Ireland. Besides these commonly researched countries, there is an extensive list of countries that are less prominent in the archival collection, yet the spatial range is still noteworthy (see Table 3.4).

It can be concluded that a vast majority of the undergraduate research emphasised an area within Scotland. Unsurprisingly, there is a shift perceptible from the 1990s onwards towards more international research and mobility. Although this development was not something that was limited to the

Table 3.3 Top 10 countries (outside of the UK) most studied

Country	Number of dissertations
Tanzania	69
Iceland	59
USA	38
Ireland	33
Canada	27
France	23
Egypt	17
Switzerland	16
Spain	12
India	10
New Zealand	10
Norway	10

Spatial contexts of student knowledge production 41

Table 3.4 Overview of other countries or areas studied at least once

Countries or areas	Number of dissertations
Germany	7
Australia; Greece	6
Hongkong; Spitsbergen	5
Nepal, South Africa, Sweden	4
Brazil, Chile, China, Cyprus, Italy, Kenya, Malawi, Peru, Russia, Singapore	3
the Atlantic Ocean, Bolivia, Cayman Islands, Ecuador, Finland, Luxembourg, Malta, Nigeria, the continent of South America, the Netherlands	2
Afghanistan, Antarctica, 'the Arctic', Bulgaria, Colombia, Congo, Czech Republic, Denmark, Dominica, Dominican Republic, Estonia, the European Union, Falklands, Haiti, Indonesia, Iraq, Jordan, Malaysia, Mexico, Mozambique, Namibia, Portugal, Puerto Rico, Saudi Arabia, Serbia, Sri Lanka, Thailand, Trinidad and Tobago, United Arabic Emirates, Zambia	1

University of Glasgow and had many economic and political causes, there was an extra impetus to do international research, even on the undergraduate level, by means of the organised fieldtrips and expeditions to countries abroad. Such curricular or extracurricular aspects of the undergraduate degree can have a long-lasting effect both in the academic careers of undergraduate students taking part and the research done within the School or Department as a whole. Unfortunately, more recently there has been a lack of such initiatives, caused by the workload of academic staff and decrease in financial support from the University.

Emergence of the microscale: biographies, bodies, and bothies

The question of 'scale' in dissertation research is obviously a spatial question but is also very strongly connected to the foci of inquiry and methodologies. In the previous sections, the global and virtual scales have already been mentioned, as well as how certain scales are more prominent in dissertations addressing certain sub-disciplines: for example, the national scale in political and cultural geography dissertations, as well as the 'regional study' as a sub-discipline with a very specific scale. From approximately 2010, other scales of geographical research emerge in the collection of undergraduate dissertations. Sharing a focus on 'the small', the situated, or the singular, such study areas are, for example, confined to one building, one household, one life, or one event.[2] The dissertations in Glasgow's archival collection which focus on this small scale often have other things in common: many of these dissertations explicitly mention ideas of 'lived' and 'embodied' experiences and are dissertations on social or cultural geography.

In the cohort of 2010, there are two dissertations that explore the theme of bothies. One of the students takes a broad approach, analysing bothy culture, whereas the other student takes one specific bothy as a starting point:

> A biography of a building will be told through the events that surround it. ... This adjusted way of seeing; a micro-scale focus valuing just one place, with many small happenings and events, and the local and personal meanings it holds does however have wider, grander effects, and not just for us humans. ... To write this dissertation a number of stories will be told: of Shenavall bothy – its materiality and reverberations throughout the landscape; of individual human and animal residents and visitors that surround the place both recent and distant; of a literal journey, my personal narrative of the 'lived-in' experience of being in this landscape for a time. ... Hopefully what will emerge is something that although very local and idiosyncratic will still be universally relevant and transferable to a host of different places – not just rural but urban or suburban.
>
> (Henderson, 2010: 2–3)

This student explores research on a microscale and explicitly mentions the wider relevance of this situated narrative. It is not only the situatedness of the research that distinguishes a dissertation such as Henderson's from many others, since there are obviously many more dissertations that stay very close to their chosen case study or study area, but also the active presence of the student-researchers themselves in the written dissertation. They are not just a 'researching instrument' whose senses are used to observe but they also start to use their own experiences, feelings, and bodies as valuable sources themselves. Such autoethnographic approaches are connected to the use of language that is highly personal, emotional, and story-like.

In her dissertation *Embodied geography of tattooing: Scratching the surface of female tattooists* (2014), Claire Sannachan explores the theme of gender and tattoos using a combination of different methods. More traditional methods, such as semi-structured interviews, are enriched with her own experience of being tattooed. In this, she combines her own experiences, but also takes 'the body' as a geographical space and place:

> To further unpack the female tattooed skin, it is vital to look at skin and the body as a whole within geography. The female experience has been under represented, both as a professional tattooist and as alteration of the female body. Along with this research, the bodies' skin will be considered as a space and place while examining the transformation of female tattooing.
>
> (Sannachan, 2014: 6)

Emphasising 'innovative' scales asks for solid justification. It is striking that all the examples mentioned here provide an extensive explanation of why and how the microscale or small scale that they use in their research deserves a place in geography's disciplinary framework.

Another example of how the microscale has been used is the emergence of biographical studies in the dissertations. Some of these dissertations are 'traditional' biographies of an individual; for instance, on an artist, while others provide a biographical study of a landscape or a practice. The idea of approaching other entities than human beings or organisations with a biographical perspective is not new. In 1984, for instance, Pred (1984: 288) already wrote 'that all humanly shaped landscape elements as well as all humanly made objects are not lifeless, not without biographies of their own that are part of the never-ending transformation of nature'. Before 2000, the undergraduate dissertation archive does not entail any examples of such biographical studies or embodied experience nor examples of student-researchers focusing on one specific small space. It thereby demonstrates some kind of a 'delay' in the emergence of such approaches that are likely to be connected by curricular changes and staff changes within the department.

Going into the field

'The field' is often positioned opposite 'the lab' as a space of knowledge production, for instance by Kuklick and Kohler (1996). The difference between the two is characterised by the contrast of the field being a 'natural field location' and the lab being a 'closed-system' wherein control and prediction becomes possible. For many undergraduate students, it was not either-or but both-and. In many physical geography projects, a period of fieldwork preceded a period of lab work, open to all manner of uncontrollable and unpredictable flows, agents, and phenomena, the first playing a role in the data *collection* and the second in data *analysis*:

> The basic field technique was the analysis of spoil tips which were randomly selected without knowledge of their type, age or characteristics. This field technique consisted of placing a transect through what was preferentially chosen as a representative portion of the spoil tip. Quadrats of one metre square were marked out at fixed regular intervals; the procedure was standardised to intervals of five metres after some experimentation to assess what would be a good sampling distance in an unfamiliar environment. Certain factors were noted directly from the quadrat; the vegetation cover, species number, slope, sub-surface temperature and environment were measured and noted. A soil pit was dug, and a field sketch of the profile made, and soil samples taken for laboratory analysis.
> (Brown, 1978: 11)

This quotation, from Kenneth Brown's dissertation entitled *An investigation into the reclamation potential of spoil tips in the Fauldhouse region* (1978) describes his research methodology, but also addresses the spatialities of his methods: his project was delimited to a specific region, and within that region, the spoil tips were randomly selected. Besides the collection of data by making use of quadrats, taking notes, and making sketches, he also took soil samples for laboratory analysis. This is a routine that is seen frequently in all kinds of physical geography dissertations over time, but from the mid-1990s the work in the lab is described and documented more elaborately. This is not only connected to the fact that methodology chapters gained a more central place in dissertations over time but was probably also influenced by the improvements of the departmental laboratory facilities around 1994. The lab improvements afforded better working facilities for dissertation research: it became possible to continue on the same project for a few days in a row instead of always checking whether the lab space was needed for teaching. Besides these improvements in the research facilities in terms of accessibility, the lab improvements also included new instruments, such as stream tables. This meant that students working on a fluvial geography dissertation were able to do more experimental research. From the mid-1990s, students regularly made use of these facilities: sometimes complementing their fieldwork but other times as their main method; and more theoretical, methodological, and experimental research projects began to appear in the collection of dissertations based *solely* on research undertaken in the laboratory:

> ... five experiments testing the Be-separation effectiveness of larger cation-exchange columns were carried out at the University of Glasgow's School of Geographical and Earth Sciences. ... The required acids, chemical rock sample substitutes and solution standards were prepared for the experiments. ... The experiments were conducted in a fume hood in a geochemical laboratory. Column stands were equipped with two plastic columns with large yellow reservoirs attached in order to hold a larger volume of eluent.
>
> (Vanik, 2014: 23–24)

It is a variety of lab facilities that students use, from the geochemistry, to the flume (see Figure 3.1).

Such lab-based research projects are nonetheless still exceptions: the lab was for physical geographers definitely a 'usual' space of knowledge production, but not regularly the *only* space. The experience of going into a field, even if combined with lab work, has hence been a consistent factor since the 1950s.

The fieldwork locations of students undertaking a human geography research project for their undergraduate dissertation are often 'open' spaces such as cities, villages, neighbourhoods, and tourist destinations. Shifts are perceptible in line with disciplinary trends. For instance, with emerging attention paid to the social geographies of outsiders (this became an Honours option course in 1997), spaces such as schools, nurseries, care homes for elderly, and refugee centres also started to become viable fieldwork locations:

Figure 3.1 Photo of the flume, located at the School of Engineering (Baff, 1994: 17).

the site [primary school] remains an interesting and viable 'laboratory' for investigating the hierarchy of possible 'places' of association in accordance to a child's emerging political identity.

(Ferguson, 2010: 11)

Such sites bring specific ethical considerations with them. In many cases, there is a personal relationship with the fieldwork location: for instance, a student's own former primary or secondary school[3] or the nursery[4] the student's child attends.

This choice demonstrates that access to certain spaces has often to do with existing social networks and relations. Just as many physical geography dissertations combine fieldwork with laboratory research, many human geography dissertations combine fieldwork with documentary analysis, such as archival research. In the regional geography dissertations from the 1950s to early 1970s, every student mentions one or more local libraries and archives located in the region that they were researching. In some cases, documentary research was the only method and libraries and archives were thus the main field work location: 'this project was principally an exercise in library field work' (Haynes, 1970: n.p.). Archives are hence also important fieldwork locations, mostly for historical geography dissertations, but also as some form of supportive fieldwork location for, for instance, urban geography projects. Spaces connected to documentary analysis and archival research are often local or regional archives

46 *Spatial contexts of student knowledge production*

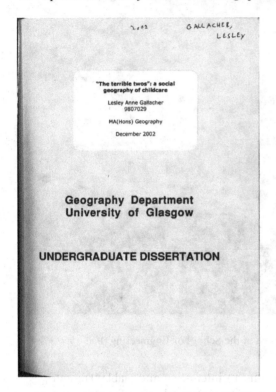

Figure 3.2 Dissertation cover (Gallacher, 2002).

and national or regional 'record offices', such as the Scottish Record Office (now named the National Archives of Scotland). This analysis of what kind of spaces actually are meant when talking about the fieldwork locations of undergraduate students demonstrates that, although the 'classic' image of geography fieldwork is prevalent and dominant within all cohorts, both in its physical geography and its human geography conception, it is important to acknowledge some changes over time. These shifts are related to changes in the discipline as well as changes in the local and departmental contexts.

The practicalities of fieldwork

Fieldwork is often seen as the 'signature' practice of geography (Day and Spronken-Smith, 2017: 2) and as the most powerful educational activity in teaching geography (France and Haigh, 2018). Fieldwork is an integral part of the geography undergraduate curriculum, but the experience of doing *independent* fieldwork for the dissertation is a different, individual experience. The activity of 'going into the field' often started with borrowing a bike, taking a train, or asking parents for a ride. However, there is usually some extensive less visible planning period that proceeds such journeys to the field. In some cases, the struggles experienced in planning and preparing greatly shape the scope, process, and output of the fieldwork. Other struggles or challenges are not apparent before the actual fieldwork has started and ask for a solution 'on the spot'. The analysis of practical struggles, as well as the more positive equivalent of expected and unexpected practical opportunities, evokes certain moral discussions about the accessibility of and equality in academic education.

Planning fieldwork starts with deciding on the scope of the research project, and sometimes making a good plan required preliminary survey work in the field (e.g. Sinclair, 1978). For the majority of the cohorts of geography students in Glasgow, dissertation research has taken place in the summer between the third and fourth year of the four-year undergraduate programme. This fixed moment to do dissertation research brought particular problems for some of the students.

Figure 3.3 Land use survey photographs (Waddell, 1974: n.p.).

For instance, Alison Waddell researched land use for her dissertation *The changing extent and importance of derelict land and land reclamation in Stoke-on-Trent* (1974) by means of air photo interpretation (pre-existing photos) and a survey of the area (see Figure 26). However, she describes the lack of activity in the photos that she has taken herself and blames this lack on the timing:

> The fact that in each case the photography had been flown during the summer holiday fortnight did not help as any visible signs of activity such as smoking chimneys, lorries on sites or cars parked beside works were absent.
> (Waddell, 1974: 15)

Indeed, Waddell's photos are in themselves visually attractive, yet the lack of activity did not help her research on land use (land use patterns could still be discerned, but there were no people around to ask for more information). Not only the timing with regard to the summer season and holidays is challenging, other students had additional planning problems:

> The timing of research made reaching potential participants very challenging – the ending of summer holidays and Ramadan and the start of schools were reasons why women interested did not have time to take part.
> (Kakela, 2014: 12)

Such examples demonstrate that the 'fixed' moment of when the dissertation should be researched is shared among many generations of students and is something that might influence the kind of research that students are able to do, as well as whose help is potentially available.

Besides deciding on the scope of research and the time and timing constraints, there are many other practicalities that shape the process of fieldwork. Many of these aspects are 'universal', in the sense that they happen to students whether doing research in the 1960s or in the 2010s, and in projects that are undertaken in Scotland as well as abroad. Challenging to many human geographers, for instance, has been the poor response rate or poor response 'quality' while looking for questionnaire or interview participants:

> The initial indication was that the employers concerned would be sufficiently co-operative to allow at least an accurate cross-section of firms to be interviewed. This indication, however, proved to be rather fallacious.
> (Weir, 1970: 2–3)

> Problems with the younger generation, almost universally young males, were that they did not take the survey seriously. This was done by making outrageous comments, abusing the system, and the police department as a whole. Thus, a few questionnaires were ruined.
> (Adam, 1998: 16)

> People do not like speaking about how much they are earning, especially when their income cannot be classed as wholly legal.
>
> (McCormack, 2010: 14)

> In multiple occasions individuals I approached were under the influence of alcohol, they were often gregarious and eager to assist; however, due to their intoxication they couldn't read, understand or complete the questionnaire. They verbally communicated their aggravations, often pugnaciously.
>
> (Johnston, 2014: 19)

The nature of the 'problem' of participants is thus diverse – from not finding people that have time for you to not getting the right information or mutinous teenagers – yet consistently perceptible throughout the different cohorts. Other fieldwork limitations and challenges are more specific in more recent times; for instance, the very justified constraints raised by the need to get formal ethics approval from the university-based ethics committee.

There are also some other common fieldwork factors that play positive and not-so-positive roles in the experiences of students, from needing certain weather conditions to do research:

> Many visits to the site proved unsuccessful due to lack of wind. However, one day in September there seemed to be enough wind to move sand grains and thus enable fieldwork to be carried out.
>
> (Hamilton, 1994: 15)

To weather ruining many days of fieldwork:

> The greatest general limitation within the study was the weather, and specifically the high rate of precipitation. ... Work was hindered by the rain, and on some occasions field work had to be abandoned.
>
> (Lemon, 1994: 14)

From handy and supportive scientific instruments:

> The instrument [Kern DKM2 Theodolite] was light and easily hand portable. Setting up was very easy once the centering leg principle had been understood and operated a few times.
>
> (Graham, 1966: 12)

To DIY solutions, bottles, and buckets:

> Forty 1 litre milk bottles were rinsed and soaked in a strong detergent for 24 hours and then rinsed thoroughly with distilled water to ensure no

detergent clung to the sides. The bucket and funnel which were used during the collection of samples were also sterilised.

(Montgomery, 1998: 10–11)

Language differences, alongside unexpected events such as strikes and transport issues, can be added to the list of practical constraints.

Assistance and help in the field

Although undergraduate dissertations are *independent* research projects, many students had some sort of support in the field as well as in the preparation and writing-up stage. This support sometimes had a more intellectual character (e.g. offering information, being a respondent in the questionnaire, referring students to interesting data or locations), and other times, it was more support in a practical or counselling sense. Some of these roles are prevalent in all cohorts: this includes the help of family, friends, and peers and also the help from local residents in the study area. There are also many institutions that are thanked by students from all different decades, such as certain city council departments and forestry commissions. The help from family members and friends through the decades can be summarised as practical and psychological support, expressed by fieldwork assistance, financial support, and much needed distraction:

> my good friend, Miss Wilma M. Young, who painstakingly translated all the German I could desire and more, especially during my 'fact-finding' trip to the Federal Republic
>
> (Birch, 1986: iii)

> My father for accompanying me on each of my bird surveys to take down notes of what I saw and heard.
>
> (Aitchison, 1998: 55)

> The author wishes to firstly thank my parents who without there [*sic*] help I could not have visited Malawi, and subsequently lived there. They allowed me to disappear into the countryside and explore what it had to offer.
>
> (James, 1998: ii)

> My Grandpa for all his humour and timely hedge cutting and gardening chores when needing to let off steam.
>
> (Macari, 2002: n.p.)

Besides such forms of help, as already mentioned, it was not uncommon for mums, girlfriends, and sisters to have helped by typing up the dissertations of their sons, boyfriends, and brothers, especially between the 1960s to early

1990s. It demonstrates a surprisingly gendered element of dissertation writing: even though the cohorts of students have always been mixed, female students tended to type their dissertations themselves or paid a typist service provider to do it, whilst many male students 'outsourced' this job to female relatives. It might be a trivial element (to what extent is the *typing up* actually part of the dissertation work, when the student has done all the intellectual work himself already?), but it addresses how the 'workload' of doing an undergraduate could be different for students, even when they were part of the same cohort of students.

Examining the mentioned support and assistance chronologically, there are some shifts perceivable over time. For instance, in the 1950s and 1960s, the most thanked 'group', by a large majority, were farmers (of specific parishes, villages, and areas). This of course directly relates to the types of research the students were doing. The dissertations suggest that farmers were not only asked about the land use of the land they owned themselves but also as residents who often had lived in this specific space all their lives. Another example of a group that often was able to help out in regional studies was religious people in parishes, such as reverends and vicars. In multiple accounts of the 1950s and 1960s, they are mentioned specifically. After these decades, religious people are also sometimes mentioned, but then only in cases where religion was the main focus of inquiry. In these cases, religious people interviewed were approached as experts on a topic or as 'research participants'.

There are also certain roles or institutions that were able to offer assistance to undergraduate dissertation students of many generations, but who, more recently, have perhaps grown 'out of the picture'. Archivists and librarians working in local or regional archives, libraries, and museums played a big role in data collection from the 1950s to the 1990s. The same goes for employees at various census offices, county councils, and district councils. From the turn of the century, such parties seemed to have disappeared from the dissertations. It might be the case that they only disappeared in their *explicit* form: whereas students up to the 1990s needed physically to visit or phone these institutions, from the 2000s it became easier to find information online. This does not mean that the work of archivists, libraries, and so on is not consulted anymore, but that, because of the ease of checking such data online, it has become more or less 'invisible' that the work on inventories and digitalisation is still undertaken by actual people.

The 1970s and 1980s demonstrate other regularly consulted sources in the field: for instance, tourists, hotel owners and shopkeepers, and tourist information offices are becoming more prominent in these decades. Again, this relates to certain sub-disciplinary trends. There is, however, also a group that became indispensable in fieldwork (especially for human geographers) from the 1970s onwards: respondents and research subjects. With methods such as questionnaires, interviews, participant observation, and ethnography, it became important to connect to people in the intended response group: this crucial shift to *actively* recruiting research subjects – from casual 'informants' (giving contextual information) to people purposively researched using specific methods.

Whereas respondents provided a very active and central role in the research itself, many students also thanked those around them that had to 'endure' their behaviour and absence:

> Mum, Dad, Lesley and Rachael for help, encouragement and patience with a certain (often frustrated !) dissertation writer.
> (Moyes, 1990: 46)

> Apologies go to my children Yvonne and Graham, for whom the word 'dissertation' is now synonymous with 'dinner is going to be late again!'
> (Caulfield, 1994: n.p.)

These social relationships that play a role in the fieldwork experiences of undergraduate students demonstrate the breadth and diversity of the social contexts of the students. It also indicates that the practice of doing research for the dissertation and the writing-up of the dissertation were inextricably intertwined with the daily lives of the students: even if friends, flatmates, or family members were not directly involved in 'assisting' the student in any way, this did not mean that they did not at least notice the dissertation activities indirectly. Whereas the division between a 'private life' and a 'student life' are perhaps in general less strict, during the dissertation research this division faded even more, especially in cases where practical in-field assistance was offered.

Field expeditions

Whereas the geography undergraduate curriculum has always included field classes of multiple days in the first three years of the four-year programme, the dissertation research fieldwork has often been an independent endeavour for students. There are some exceptions, however, in the form of field expeditions: both expeditions which were organised for the purpose of undergraduate dissertation research and expeditions with a more diverse travel party. In 1966, a small number of undergraduate students joined lecturers and postdoctoral researchers to the Island of Rùm. This expedition was aimed at surveying parts of the island, on behalf of the Nature Conservancy (Graham, 1966). Two dissertation students were involved in the surveying as part of their dissertation research: two other undergraduate students joined the party but with their own independent research aims within the field of biogeography. Their dissertations reflect on sharing accommodation and transport to the island with the others, but do not indicate a lot of shared field activities. This aspect is different in other, later field expeditions, notably the multiple expeditions to Iceland in the 1990s, organised by David Evans. These trips included many more students. The four weeks shared over summer obviously meant that stronger relationships were built with peers, leading sometimes to personal acknowledgements:

> Jenny Pearce – for those unforgettable moments we spent each day crouching behind boulders, discussing the merits of oatcakes and the universe.
>
> (Partington, 1994: 3)

> everyone involved with the Iceland 94 expedition, particularly those diamond geezers with the levelling data: Ian and Gary. A big boo-yah to Morg and Spider, and the rest for making the four weeks there a lifelong experience (dig those rainy Winchester Club days man!).
>
> (Wilson, 1994: 48)

In the interviews held with current and former staff members, such field expeditions were also discussed in a different way. Philo reflects on the idea of 'unfairness' that some students held if *not* going on field expeditions themselves, explaining why staff just being 'there' on an expedition does not automatically mean that they had more input than when a student doing a non-expedition project would have consulted staff member regularly as a 'resource':

> There was disquiet for instance about students that went on Dave's expeditions, or who went with John and Jo to Africa, that they were getting a 'Rolls Royce treatment'. I never really bought into that too much. I think that – I have always tried to think of that as – a staff member is a resource. And students should feel free to draw upon that resources. Some students draw upon it more than others.
>
> (C. Philo, interview, 2019)

John Briggs, himself an organiser of field expeditions to Egypt as well as Tanzania in the late 1990s and 2000s, reflected on the quality of dissertations based on these field expeditions. He argues that there are two reasons for the high quality of dissertations connected to research expeditions: overall, the 'better students' were more interested to go on such research trips and there was some 'sense of competition' in these groups. 'When going abroad together, you do not want to be the one "doing the least"' (J. Briggs, interview, 2018).

The field expeditions meant building close connections to peers and staff members, but also led to stronger relationships as well as growing interest in specific subjects or sub-disciplines. In one of the expeditions to Tanzania, organised by Briggs and Sharp, students from Glasgow collaborated with students from the University of Dar es Salaam. For the latter, helping with translation was a very practical way of offering field assistance and overcoming language and culture barriers. These field expeditions strongly influenced students on a personal level:

54 *Spatial contexts of student knowledge production*

Figure 3.4 Cover of dissertation based on a field expedition to Tanzania (Scholes, 2010).

On returning to Glasgow after spending 3 weeks in Dar es Salaam, I remembered the culture, the people and the beautiful landscapes all of which seemed illustrative of the process of migration in one way or another. Whilst on safari immediately after completing the research I was still consumed by the concepts and realties of rural-urban migration, and witnessing the fascinating sights of hundreds of wildebeest migrating across the Ngorongoro Crater highlighted the simplicity and natural qualities of migration.

(Hammond, 2010: 6)

It is important to note that such expeditions were not only influential on a personal level, but also affected future career paths of undergraduate students. There are multiple examples of students who went to Tanzania and continued with PhD research, even with a focus on Tanzania or development geography. It demonstrates that, although the undergraduate dissertation is a relatively small exercise in research, it can be experienced by individual students as a key moment or starting point for later developments in life.

Conclusion

The conceptual pluralist understanding of 'the field' as a space of geographical knowledge production is supported by the many different fields to which students have travelled and have described in their undergraduate dissertations. From a dance school to glacier 'snouts' and from a chicken coop to machairs; the geographical field is definitely diverse. This explosion of field possibilities also influences the variety of practical challenges encountered in the field. Some students rely heavily on family or friends who act as unpaid research assistants, maybe indicating how students with a 'richer' social network might have a more pleasant fieldwork experience with a lower workload. Apart from the 'wealth' in the social sense, socioeconomic factors play a role in different opportunities for fieldwork: some students mention their parents as 'funders'.

The educational context of the undergraduate dissertation as part of a wider curriculum distinguishes such a research project from other academic research. It is not only the awareness of students and supervisors that the dissertation should be 'good enough' to pass (or to get a certain aspired grade) but also the relationship between what the students had learnt and which fields the students had come to know well during their coursework that in many cases directly influences the dissertation. Specific courses, individual staff members, and existing facilities within the department or university might influence the choices made by students. These aspects mean that, whereas many undergraduate geography students in the UK might research similar phenomena in similar 'fields', there will always be substantive differences in when certain disciplinary trends become visible in undergraduate dissertations or which areas are more regularly studied. The 'where' of dissertation research is multi-layered and diverse. The small knowledge productions in the dissertation archive all include quite specific geographies of how the research for the project is undertaken and experienced; yet, taken together, this collection has revealed numerous key geographical, social, and educational constants and changes over time.

Intermezzo 3: Rural Depopulation

In many of the regional studies of the 1960s and 1970s, the topic of rural depopulation plays a central role in the analysis of the region chosen as the study area. For instance, Paul Cortopassi explores a Scottish region in his regional study entitled *The Parishes of Elgin, Birnie, and St. Andrew's Lhanbryd in the County of Moray* (1966):

> ... the problem of rural depopulation is as serious here as elsewhere, and it may be that Government influence be brought to promote industry here; and if so this would most likely be some form of agricultural industry of the 'Baxter's' type – meat packing, or perhaps fish canning, but the possibility is a remote one. This is an agricultural region with the advantage of having a centre in Elgin to supply sufficient work to absorb most of the young people leaving the farms, and there should be no complaints if things are left as such and allowed to progress and expand in the fullness of time.
>
> (Cortopassi, 1966: 20)

Dickinson, mentioned already in Box 2.1, also analysed rural depopulation in his regional study *The geography of the Helmsley District* (1966):

> Thus Helmsley has advanced to its present position, a town of around 1200 people, but at a cross roads in its development. The three developments mentioned have helped to arrest the economic decline and depopulation which had been going on for 300 years, but problems remain. As mentioned already, in several of the parishes of the region depopulation is still severe, and in Helmsley changes in age structure are notable.
>
> (Dickinson, 1966: 24)

A later dissertation about rural depopulation, written in 1982 by Allan McMinn, examined the perceptions of secondary school pupils on their plans for the future and its relationship to the area where they grew up. In his dissertation, *Perception of migration by pupils in fourth, fifth and sixth year attending Islay High School* (1982), he explicitly delimited rural depopulation as his focus of inquiry. However, his approach, methodology, and the research questions asked were very different from the regional projects from the previous two decades that had also emphasised rural depopulation:

> The fundamental aim of the study was to determine the basic migration intentions of pupils attending Islay High School who were nearing the end of their school careers and to determine the main reasons behind those personal decisions.
>
> (McMinn, 1982: 7)

McMinn's approach not only differs from the regional dissertations because there is a more defined focus of inquiry but also by the concern for 'intentions', 'reasons', and 'personal decisions', indicating an interest in more qualitative dimensions of how people interact with a phenomenon instead of just collecting more quantitative 'facts' of the matter.

The change in language, expressed by this difference between McMinn's perception study and the earlier regional studies, is also caused by changes in what dissertations actually 'do' or encompass. McMinn, for instance, also reflected on the sub-discipline of rural geography compared to an allegedly more 'popular' urban geography, voicing a worry often heard in the academic literature of rural geography at that time:

> With the rapid and sophisticated growth of research and expertise in aspects of urban geography, the field of rural geography has been relegated to an inferior position. However, important functional changes are occurring in the countryside as traditional rural activities such as agriculture release workers from local employment.
>
> (McMinn, 1982: 4)

The central theme of 'rural depopulation' has been not only approached from a demographic perspective but also connected to very central questions about economy, industry, and employment. In later years, specific rural social issues are still researched by some students, although the number of rural dissertations has become significantly lower than urban geography dissertations. Many of the rural geographies undertaken by students since the 1990s focus on the

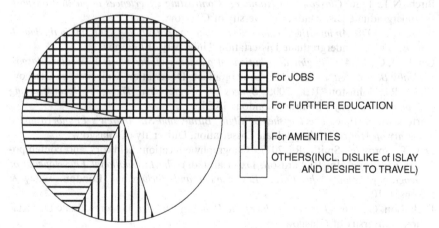

Figure 3.5 Graphs of questionnaire results, displaying 'major reasons why people want to leave Islay' (McMinn, 1982).

sense of community in a rural area or smaller towns, often emphasising the relationships between 'locals' and 'newcomers' in towns, rural economies, and countryside tourism.

Notes

1. A library study of course still takes place *somewhere*, but the difference I am making here is between travelling to the area of study (and maybe also visiting some archives and libraries *in situ*) and, for instance, studying a foreign region/area based on research undertaken in the University Library in Glasgow.
2. Physical geography dissertations have often expressed an emphasis on a singular element of the landscape – one river, one glacier, or one beach – long before 2010 but I have interpreted the 'singularity' of a river as similar to the 'singularity' of, for instance, one neighbourhood in a city or as a study of one population group ('the elderly' or 'adolescents'), which is also present in the entire temporal scope of the dissertation archive.
3. "Permission was granted by the Head Teacher because I am a former pupil of the school and my old geography teacher is an Assistant Head" (Adamson, 1998: 6).
4. "Perhaps the most important practical issue concerning this research was the presence of my son, Aidan, in Stepping Stones. This influenced all aspects of the research and my research role in particular" (Gallacher, 2002: 19).

Bibliography

Adam, B.J., 1998. *Do CCTV cameras in Strathclyde reduce crime and the fear of crime?* Undergraduate Dissertation, University of Glasgow.

Aitchison, F.C., 1998, *Woodland management on Loch Lomondside*. Undergraduate Dissertation, University of Glasgow.

Baff, D.G., 1994. *The effect of sediment sorting on the development of bedforms: A flume study using medium sands.* Undergraduate Dissertation, University of Glasgow.

Birch, N.T., 1986. *Glasgow and Hamburg: Contrasting experiences in public transport.* Undergraduate Dissertation, University of Glasgow.

Brown, K., 1978. *An investigation into the reclamation potential of spoil tips in the Fauldhouse region.* Undergraduate Dissertation, University of Glasgow.

Caulfield, C., 1994. *Changing distributions of pulmonary tuberculosis in Greater Glasgow Health Board Area, 1974 to 1993.* Undergraduate Dissertation, University of Glasgow.

Cloke, P.J., Johnston, R.J., 2005. *Spaces of Geographical Thought: Deconstructing Human Geography's Binaries.* London: SAGE.

Cortopassi, P., 1966. *The Parishes of Elgin, Birnie, and St. Andrew's Lhanbryd in the County of Moray.* Undergraduate Dissertation, University of Glasgow.

Day, T., Spronken-Smith, R., 2017. 'Geography education: fieldwork and contemporary pedagogy'. In: Richardson, D., et al. (eds) *The International Encyclopedia of Geography: People, the Earth, Environment and Technology.* London: Wiley & Sons, 1–10.

Dickinson, G., 1966. *The geography of the Helmsley District.* Undergraduate Dissertation, University of Glasgow.

European Union, 2012. *Erasmus: changing lives, opening minds for 25 years.* Luxembourg: Publications Office of the European Union.

Ferguson, N., 2010. *Children's' perceptions of identity in the Scottish Highlands: Fledgling political notions and a sense of belonging.* Undergraduate Dissertation, University of Glasgow.

France, D., Haigh, M., 2018. 'Fieldwork@40: fieldwork in geography higher education'. *Journal of Geography in Higher Education*, 42(4), 498–514.

Gallacher, L.A., 2002. *"The Terrible Two's": A social geography of childcare*. Undergraduate Dissertation, University of Glasgow.

Graham, N.M., 1966. *A survey to establish ground control for aerial photographs covering the north-west of the Island of Rhum*. Undergraduate Dissertation, University of Glasgow.

Hamilton, F.J., 1994. *The effect of dune vegetation on wind velocity and sand transport across a sand dune system*. Undergraduate Dissertation, University of Glasgow.

Hammond, C., 2010. *City strangers: An investigation into rural-urban migration, Dar es Salaam, Tanzania*. Undergraduate Dissertation, University of Glasgow.

Haynes, A., 1970. *Effects of the Agricultural and Industrial Revolutions on the Parishes of Hamilton, Dalserf, Glassford and Stonehouse*. Undergraduate Dissertation, University of Glasgow.

Henderson, S., 2010. *Shenavall Bothy*. Undergraduate Dissertation, University of Glasgow.

James, M., 1998. *The consequences of present transport infrastructure on trade; Malawi*. Undergraduate Dissertation, University of Glasgow.

Johnston, F., 2014. *Fear of crime in Glasgow - A study of the Subway vicinities*. Undergraduate Dissertation, University of Glasgow.

Kakela, E., 2014. *Somali diaspora in Helsinki: Negotiating gender and religious identities in a secular space*. Undergraduate Dissertation, University of Glasgow.

Kohler, R.E., 2002. *Landscapes & Labscapes: exploring the lab-field border in biology*. Chicago: University of Chicago Press.

Kuklick, H., Kohler, R.E., 1996. *Science in the Field*. Chicago: University of Chicago Press.

Leith, M.S., & Soule, D.P.J., 2011. *Political Discourse and National Identity in Scotland*. Edinburgh: Edinburgh University Press.

Lemon, K.M., 1994. *An investigation into lateral and termino-lateral deposition during New Zealand's Otiran Glaciation*. Undergraduate Dissertation, University of Glasgow.

Livingstone, D.N., 1995. 'The spaces of knowledge: Contributions towards a historical geography of science'. *Environment and Planning D: Society and Space*, 13, 5–34.

Macari, P., 2002. *A social geography of Glasgow's west end's Pakistani Muslim community: Has sense of place changed post September 11th 2001?* Undergraduate Dissertation, University of Glasgow.

McCook, S., 1996. '"It may be truth, but it is not evidence": Paul du Chaillu and the legitimation of evidence in the field sciences'. *Osiris*, 11(1), 177–197.

McCormack, K., 2010. A study of livelihood: Kunduchi Village, Dar es Salaam, Tanzania. Undergraduate Dissertation, University of Glasgow.

McMinn, A., 1982. *Perception of migration by pupils in fourth, fifth and sixth year attending Islay High School*. Undergraduate Dissertation, University of Glasgow.

Mitchell, J., 2014. *The Scottish Question*. Oxford: Oxford Scholarship Online.

Montgomery, F.C., 1998. *Effect of human activity on physical and chemical water quality in an upper section of the River Spey, and assessment of human reaction to pollution issues*. Undergraduate Dissertation, University of Glasgow.

Moyes, G.D., 1990. *Glasgow's trams: The past and the future*. Undergraduate Dissertation, University of Glasgow.

Partington, E.A., 1994. *The comparison and analysis of two methods of collating ablation data for the snout of Sandfellsjokull, Myrdalsjokull Ice-cap, southern Iceland.* Undergraduate Dissertation, University of Glasgow.

Pred, A., 1984. 'Place as historically contingent process: Structuration and the time-geography of becoming places'. *Annals of the Association of American Geographers*, 74(2), 279–297.

Sannachan, C., 2014. *Embodied geography of tattooing: Scratching the surface of female tattooists.* Undergraduate Dissertation, University of Glasgow.

Scholes, S., 2010. *The impact of disease on the socio-economic structure of the household in Dar es Salaam, Tanzania.* Undergraduate Dissertation, University of Glasgow.

Sinclair, M., 1978. *The glacial geomorphology of the middle Endrick Valley - West central Scotland.* Undergraduate Dissertation, University of Glasgow.

Teichler, U., 2009. 'Internationalisation of higher education: European experiences'. *Asia Pacific Education Review*, 10, 93–106.

Thompson, I.B., Dickinson, G., Lowder, S., Paddison, R., 2009. 'Recollections and reflections'. *Scottish Geographical Journal*, 125(3–4), 329–343.

Vanik, P., 2014. *Testing ion exchange column efficiency for titanium, beryllium and aluminium separation.* Undergraduate Dissertation, University of Glasgow.

Vetter, J. (ed), 2011. *Knowing Global Environments: New Historical Perspectives on the Field Sciences.* New Brunswick: Rutgers University Press.

Waddell, A., 1974. *The changing extent and importance of derelict land and land reclamation in Stoke-on-Trent.* Undergraduate Dissertation, University of Glasgow.

Weir, J.C., 1970. *Towards a further understanding of commuting patterns in the new town of East Kilbride.* Undergraduate Dissertation, University of Glasgow.

Wilson, D.J.H., 1994. *The application of lichenometry and palaeodischarge techniques into the interpretation of the glacier foreland of Sandfellsjokull.* Undergraduate Dissertation, University of Glasgow.

4 Becoming a geographer
Dissertations as intellectual source material

Undergraduate dissertations contain individual narratives about research projects undertaken using specific skills, methods, and tools. Investigating these dissertations as social and cultural sources provides insights into both the role of fieldwork in the geography curriculum and the active research experiences of many geographers-in-the-making. Each dissertation is also an *intellectual* source, however, a piece of writing, approximately between 6,000 and 10,000 words long (the length of dissertations at Glasgow University has increased slowly over the years), with a specific aim, hypothesis, question, or objective. Each dissertation is an intellectual encounter, the deployment of an academic awareness and capacity to inquire into a chosen substantive subject matter, a topic for study both suggested by the 'real world' and, if sometimes awkwardly, aligned with the more systematic or subdisciplinary interests of the geographical academy. In this latter respect, the dissertation archive as a whole discloses disciplinary trends and shifts: to do with the disciplinary divide between human and physical foci, of course, but also from subdisciplines maybe going extinct at a slow pace through to more ephemeral disciplinary 'hypes'. The collection of Glasgow's geography dissertations also discloses striking continuities throughout the decades. In this chapter, such disciplinary shifts and continuities will be explored by an analysis of both the intellectual content of the dissertation archive as a whole and some specific dissertations in particular. First, the 'holistic' approach of the geography curriculum, including both human and physical geography, will be discussed. This form of academic geography education is not a global trend or development. In many non-Anglophone higher education contexts, such as at Dutch universities, undergraduate students either follow a human geography degree programme or a physical geography or earth sciences degree programme. This unity within the British geography curriculum leads to certain bridges between human and physical approaches, providing some kind of 'hybrid' geographical knowledge. This chapter displays some recent bridges between the two: for instance, hybrid studies on sustainability, climate change, and green energy. Second, this chapter discusses subdisciplinary shifts and trends over time: some quick rises of certain popular subdisciplines in the dissertation collection are connected to potential causes of such rapid shifts, focusing particularly on changes in geomorphological studies over time and the explosion of cultural and social geography towards the end of the twentieth century. This chapter, thus, explores the richness of geographical knowledge that has been produced by many generations of undergraduate students at the University of Glasgow, teasing out significant variations in 'what' has been the substantive, maybe subdisciplinary, focus of this knowledge. Students' independent research projects take place on a level between formal academic geography and non-specialist geography, and offer something distinctive to the existing narratives of what geography, or geographical knowledge, is and has been.

One undergraduate degree with two distinguished pillars

The educational context of the undergraduate geography curriculum at Glasgow University comprises both human and physical geography, leading to a degree with methods, writing styles, and foci of inquiry deriving from very different intellectual traditions. Whereas in some other countries students at undergraduate or Bachelor's level choose between human geography or physical geography beforehand, most British universities provide a 'singular' Geography degree. This unity in the educational programme is, however, not always experienced by everyone, whether students or staff members, involved. This sense of division is perhaps also encouraged by the organisation of the degree: for instance, second-year geography students at Glasgow University nowadays will take tutorials throughout both semesters, but these are taught alternatingly by their 'human geography tutor' and their 'physical geography tutor'. By dividing sessions in human geography or physical geography seminars, the presupposed distinction between the two pillars within the undergraduate curriculum might be exacerbated. Nonetheless, there are enough elements in the degree that emphasise the interconnectedness of the human and the physical, both in field classes and in the 'day-to-day' curriculum: for instance, in lectures and tutorials about the urban environment. Some foci of inquiry provide a more obvious connection between the two pillars. Examples are regional geography dissertations, comprising divergent social and physical aspects of a region, and geographical studies of hazards, a more recent 'bridge' between human and physical geography, with projects on, for instance, flood risk and flood risk perception.

Figure 4.1 addresses the absolute number of dissertations categorised as human geography and physical geography, or as 'both', suggesting an integral, more hybrid form of geography combining the human and physical. Of course, this categorisation exercise is somewhat arbitrary, but it is still insightful in many ways.

Figure 4.1 demonstrates how human geography has always – with the exceptions of the cohorts of 1966 and 1969 – dominated physical geography in the number of dissertations per cohort. It gives rise to some general comments on the shifting ratio between human and physical geography: first, 'unified', integral geography seemed to have been dominant until the mid-1970s and after that has been marginally present in every cohort. There seems to be a small

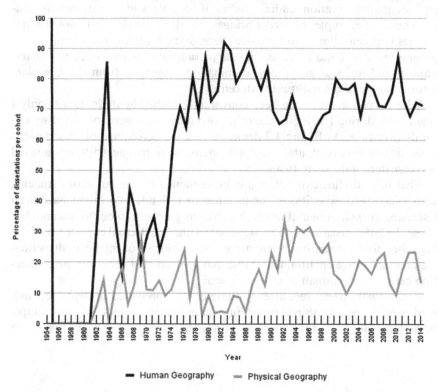

Figure 4.1 Number of dissertations per cohort that can be categorised as human geography, physical geography, or 'both'.

increase in such hybrid dissertations since approximately 2000. Although the increase over time is very small, there might be some disciplinary trends that can be identified as causes for this small increase. Second, Figure 4.1 indicates that the dominance of human geography over physical geography was strongest during the 1980s. In some cohorts in this decade approximately nine out of ten students wrote a human geography dissertation. It demonstrates the quite worrying state of physical geography in the local context of Glasgow's geography department during the 1980s. The third remark is a logical follow-up of the previous remark, namely the resurgence of physical geography in the 1990s. These three observations about the quantitative data are the basis for the following sections about the twin pillars of human and physical geography.

Recent bridges between human and physical geography

Some subdisciplines are very strongly positioned as *either* human geography or physical geography, whereas other subdisciplines can be placed closer to the

64 Becoming a geographer

middle on a hypothetical 'human to physical' scale. Examples of such subdisciplines are conservation studies, studies of hazards, and environmental studies. These are examples of *recent* bridges, but these are definitely not the first bridges between human and physical geography. Earlier examples of hybrid geographies are found in the disciplinary 'traditions' of environmental determinism and regional geography, dominating geography from the late-nineteenth century to the mid-twentieth century.

Whereas conservation has been consistently studied by students with only a small peak during the 1980s, the other two mentioned here increased significantly over time. As Figure 4.2 demonstrates, the percentage of dissertations consisting of environmental geography increased over time, with only a small decrease during the early 1990s.

That brief decline can potentially be explained by the numerous students who joined field expeditions, for instance to Iceland, who might have been interested in environmental research as well as geomorphological research but chose the latter because of the unique social and intellectual experience of the organised field expedition. There are several examples of intellectually valuable and innovative environmental research projects that focus on the connection between 'the human' and 'the physical'.

In the 1980s, there were already students who presented in-depth research studies on the relationship between humans and the environment, one example being John McAuley in his dissertation entitled *The impact of man on the water*

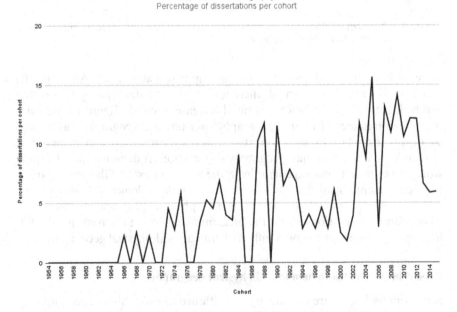

Figure 4.2 Percentage of environmental geography dissertations per cohort.

quality within the South Calder Basin (1982). McAuley describes his own research explicitly as physical geography, presenting the statistical data, maps, and information that he used concerning agricultural, urban, and industrial influences on the water quality as 'merely' secondary data. However, his extensive introductory remarks may suggest otherwise and strongly emphasise the human elements of his research:

> After two million years of exploiting nature, man [sic] finds himself in a world closely moulded to his needs and desires. Only now is he becoming aware of the consequences of his prodigality – A point... has been reached in history when man must shape his activities throughout the world with a more prudent care for their environmental consequences. Through ignorance or indifference, man can do massive and irreversible damage to the biosphere on which his life and well-being depend. An ever increasing number of interested parties are drawing the attention of Governments and peoples to the growing evidence of man-made harm in many regions, including dangerous levels of pollution in water, air, earth and living organisms, major and undesirable disturbances of the ecological balance of the biosphere, and the destruction and depletion of irreplaceable resources.
>
> (McAuley, 1982: 7)

The archaic gendered use of the concept of 'man' to describe humans distracts from the more topical conscience of the 'irreversible' negative influence of people on the (natural) environment. McAuley thus presents an example of ethically charged research that he himself describes as physical geography, but which definitely has a strong human component to it.Such ethical–political studies connecting the human and the physical appear irregularly throughout the decades from the 1980s onwards, but increase markedly for the cohorts of the twenty-first century: sustainability, climate change, and green energy are indeed three 'buzzwords' in many of these recent dissertations. In the cohort of 2010, two students independently connect the context of the environment and climate change to the role of the media. In her dissertation entitled *The role and perception of environmental journalistic cartography: Eyjafjallajokull volcanic eruption case study* (2010), Lauren Scott examines environmental cartography by analysing the maps accompanying news stories, approaching these maps as 'vessels of power' (Scott, 2010: 25). Her dissertation thus focuses on the role of media, and particularly on the role of maps in printed media, but she studies these maps from the perspective of *how* people are informed about environmental issues, thus majoring on the circulation of geographical knowledge about the environment rather than on the environmental issues themselves. The second example of such a connection between media studies and environment is Suzanne Smith's dissertation *Shaping public perceptions on climate change through the printed media: Comparing* The Guardian *and* Evening Times (2010), comparing articles about climate change from a national and regional newspaper, providing a similar study although not focusing on maps in particular.

As Figure 4.2 demonstrates, the increase in environmental geography dissertations was a gradual one. The appearance of the field of hazards studies is, however, more abrupt. With just a few dissertations on hazards during the 1990s and 2000s, the cohort of 2014 suddenly presents a peak of 10% of the students writing their dissertation on hazards. These dissertations all have a very similar focus – they are about flood risk, flood risk perception, and flood risk prevention – although, interestingly, they have been supervised by different members of staff, including historical geographers, geomorphologists, and coastal geographers. The study areas of the projects are diverse as well, from Somerset in England to Dumfries and Galloway and Aberdeenshire in Scotland, but their methodologies and approaches are very similar: all include the design and use of questionnaires that provided the opportunity for some statistical analysis as well as interviews with experts. This sudden peak in dissertations on one specific topic could be instigated by two factors and it is probably a combination of both: first, specific attention to this research focus in the undergraduate curriculum, for instance, a course on hazards taught by Rhian Thomas being of interest to many students, and second, actual and urgent attention to the theme in wider society generally and the media specifically.

Besides environmental geography and hazard studies, there is a third subdiscipline that provides an evident and consistent 'bridge' between human and physical geography, namely, conservation studies: a small subdiscipline, not subject to any significant increase or decrease over time. In the 1970s and 1980s, conservation study dissertations were often connected to the impact of outdoor recreation and tourism, and many of them took Loch Lomond as their study area: a clear example of the influence of particular academic members of staff on dissertation subjects, in this case Gordon Dickinson, as well as the role of considerations concerning money and travel in the decision about fieldwork location. It is an example of an element of geography that has 'always been there', at least throughout the second half of the twentieth century. This definitely does not apply to all foci of inquiry and subdisciplines. The examples mentioned here demonstrate that 'the physical environment' as treated in the dissertations has never been solely physical and has always been recognised as something influenced by and influencing human actors and human lives. Envisaging a two-way, mutually causal relationship between the 'natural' environment and its 'human' overlay is different from stricter conceptions of this relationship, for instance, early-twentieth century environmental determinism, arguing for a one-way arrow from the environment to the human (Philo and Ernste, 2009). Overall, there is an upward trend noticeable towards a more critical, ethical–political voice in such environmental dissertations, indicating that some of the fields traditionally closer to the epistemology and methodologies of natural sciences are also influenced by discussions taking place in social sciences and the humanities. The forging of a new environmental geography hence recentralises human–environment relationships in a way that includes factors external to 'just' the environment or 'merely' concerned with human agency (such as economic logics).

Subdisciplinary shifts

Figure 4.3 demonstrates the overall numbers of dissertations per subdisciplines. As the figure shows, there are four subdisciplines that really stand out: social, regional, economic, and urban geography. These subdisciplines are all close to encompassing 10% of the over 2,600 dissertations in the archive (urban geography around 9% and social geography 12.5%). It is possible to distinguish some of the subdisciplines that are not as prevalent as the top four but are still being studied by relatively high numbers of students, as well as some subdisciplines that have only been studied by a few individuals throughout the decades. It is interesting to note that, for instance, the number of geomorphology dissertations – as the study of landforms, their forms, and maybe processes – appears very low, with only a few dissertations allocated to this subdiscipline; but this is because many geomorphological studies adopt a more particular focus on specific landforms or formative elements, typically given names such as glacial, coastal, or fluvial geomorphology, and are thus represented in a more specific 'category'.

Table 4.1 shows the 'peak years' of subdisciplines and demonstrates some additional insights that Figure 4.3 cannot offer. Some subdisciplines have surprisingly high relative numbers, such as biogeography (1966) and economic (1968) and urban geography (1963). These numbers are likely indicative of the Honours options that the undergraduate curriculum offered in the 1960s: there was just a small number of options and students were required to write their 'systematic' dissertation for one of these Honours options.

Some other peaks can be explained by the organised field expeditions discussed earlier: for instance, the numbers for glacial geomorphology (1995) and development geography (2011). Besides these two causes, the limited number of Honours options in earlier years and the field expeditions, there are probably three other explanations of the data displayed here. First, regional geography[1] was mandatory in some of the earlier decades, leading to some '100%' numbers in a few cohorts. Second, some subdisciplines do not have a very convincing highest number and 'have just always been there' on a relatively small scale, yet across multiple decades. This goes, for instance, for rural geography and population studies. A further bundle of explanations concerns the role of disciplinary trends, the role of individual staff members in motivating and inspiring students to do certain kinds of geographical research, and the expectations and ambitions of generations of geography students. These connections here do not work in just one way, but rather in several ways, which make it hard to distinguish what exactly comes first. The next sections of this chapter will explore this complex context of (sub)disciplinary trends within one educational context in some more detail, discussing the changes in geomorphological research by students and the rise of social and cultural geography throughout the decades.

68 *Becoming a geographer*

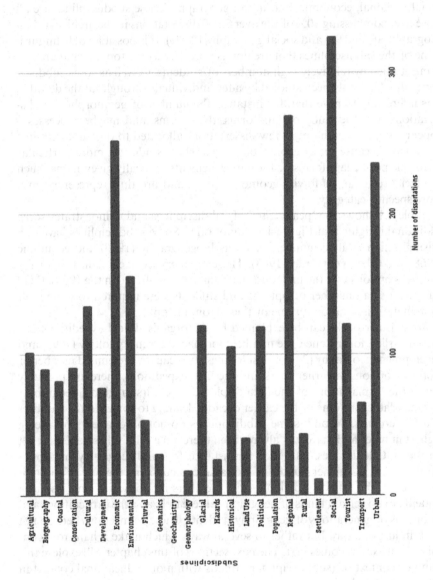

Figure 4.3 Number of dissertations per subdiscipline.

Table 4.1 Subdisciplines: cohort in which they were most popular

Subdiscipline	Cohort(s)	Percentage of cohort
Regional	1958–1961	100
Urban	1963	43
Transport	1964	18
Biogeography	1966	21
Geomatics	1967	16
Economic	1968	26
Settlement	1974	6
Glacial	1977	6
Historical	1977	28
Land Use	1979	11
Agricultural	1986	15
Population	1986	8
Political	1988	15
Conservation	1991	17
Coastal	1996	17
Tourist	1996	14
Rural	1997	5
Cultural	2003	22
Environmental	2005	16
Social	2006	28
Fluvial	2007	7
Development	2011	19
Hazards	2014	19

Geomorphology: describing landscapes, modelling landscapes, or explaining landscapes?

As Figure 4.1 has illustrated, the ratio between human geography and physical geography changed throughout the decades. Whereas the percentage of physical geography dissertations fluctuated between 10% and 30% during the late 1960s and 1970s, it dipped below 10% during the 1980s. The early 1990s saw a steady increase again, with a few of the cohorts in the early and mid-1990s being higher than 30%. These shifts are related to the already discussed field expeditions, while specific (sub)disciplinary trends, the impact of individual staff members, and educational innovations and curricular changes play important roles as well. There is a distinction between subdisciplines such as fluvial and glacial geomorphology, and 'broader scale' geomorphology, without further specification. Dissertations in this latter category are a small number of dissertations from the 1960s and 1970s that cover a larger area.[2] Of course, such studies still discuss specific types of geomorphology (dependent on the area chosen), but these do not have a specific research question or objectives focusing on either coastal, fluvial, or glacial morphology and are, to some extent, comparable to the regional dissertations of the same era: essays describing the geomorphology of a region, collecting all kinds of data, but without very specific inquiries into causes or consequences. Geomorphological projects, then, asked for specific skills of undergraduate students: *describing* a landscape is a very different thing than *modelling* a landscape, let alone trying to *explain* a landscape. One example of such a descriptive geomorphological dissertation was written by John Rankin in 1962. In his dissertation entitled *Morphology of the Ardoch Estate Area with special reference to Drift + Raised beaches*, Rankin demonstrated his skills at interpreting the landscape before him – as he undertook field visits – as well as interpreting existing maps of the same landscape:

> One of the principal difficulties in interpreting the morphological map is that of assessing whether or not flats on adjacent or widely separated spurs, and at slightly different altitudes, are in fact, the same feature. Similarly in the case of breaks of slope, where mergence or bifurcation may make it very difficult to ascertain which break, if any, is continuous with another.
>
> (Rankin, 1962: 7)

Rankin's dissertation provides a neat demonstration of analytic skills, but it is even more interesting that he includes discussions about what was difficult and what was manageable in the limited time frame and what potential follow-up projects could tackle. With a chapter dedicated to 'difficulties in analysis', he describes how he had to 'completely traverse the area on foot' and how he combined field observations, general map interpretation skills, and the consultation of geology maps (Rankin, 1962: 13). Other than his peers, Rankin mentions two 'tutors' as advisors: "Mr. H.A. Moisley of the Department of Geography and Dr. W.W. Bishop, of the Department of Geology" (Rankin, 1962: 34). As this dissertation, written in 1962, implies, physical geography projects have sometimes been closely connected with Geography's departmental sibling in Glasgow: Geology.[3] The division line is not always evident, but the existence of this other undergraduate degree at Glasgow might be an explanation of why human geography is more 'popular' as a field for dissertation research than physical geography: many students interested in the formation of landscapes, who are perhaps not *that* interested in the human side at all, have already chosen to study geology/earth sciences instead.

From the three more specific categories of coastal, fluvial, and glacial geomorphology, glacial studies have been the most popular over time, whereas fluvial geography has always been a significantly smaller subdiscipline in Glasgow. There are two peaks discernible in the dissertation archive regarding glacial geomorphology, as depicted in Figure 4.4, with the first dissertation focusing solely on glaciers and glacial geomorphology appearing during the 1970s and a second peak during the 1990s.

Dissertations on glacial geomorphology can be distinguished into two categories: dissertations which aim to provide a map, a survey, or a description of a (formally) glacial landscape and dissertations that approach a specific area with a specific, well-defined research question; often, for instance, a

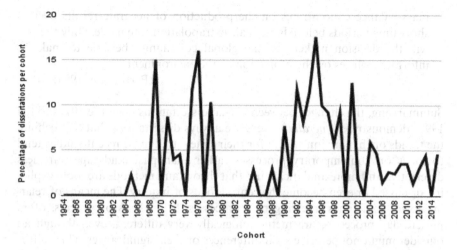

Figure 4.4 Percentage of glacial geomorphology dissertations per cohort.

contemporary glaciated landscape with an interest in *current* processes. Comparing several dissertations on glacial geomorphology from different decades demonstrates a few changes over time, not only in methods but also in the focus of research. Generally, there is a shift perceptible from formerly glaciated study areas to case studies with active ice. This is also described in (scarce) histories of geomorphology (e.g. Tinkler, 1985), with this shift starting to happen around 1960. Educational practice and the undergraduate dissertations arguably reveal a 'delay' in following such subdisciplinary trends by more than 15 years. In *The glacial geomorphology of the middle Endrick Valley – West central Scotland* (1978), Martin Sinclair aimed to draw up the glacial history of the area 'within a regional framework' (Sinclair, 1978: 1). He mapped the region as well as interpreted the forms of glacial drift deposits in the region. He provided an insightful narrative of the glacial history and changing morphology of the Endrick Valley, but also furnished some suggestions for future research. Comparing this dissertation with Elisabeth Partington's glacial research undertaken as part of the 1994 Iceland Expedition, the differences are evident and these differences are exemplary for the cohorts of which they were a part. In *The comparison and analysis of two methods of collating ablation data for the snout of Sandfellsjokull, Myrdalsjokull Ice-cap, southern Iceland* (1994), Partington (1994) undertook an essentially methodological study of doing glacial research, in a presently glaciated area, comparing two methods of estimating how much a glacier is ablating over a short period of time and seeking to judge the accuracy and practicality of these different methods. She paid attention to the *reason* why this methodological emphasis might be relevant:

> In an era of concern for the environment and postulations about 'Global Warming' it is of paramount importance that glacial research into

mass-balance concentrates on the production of accurate results over short time periods before large scale extrapolations are made. Only then will the decision makers of the global community be able to make informed policies on the protection of the environment.

(Partington, 1994: 39)

Summarising, the changes between glacial dissertations from the 1970s to the 1990s demonstrate *generally* – there are always dissertations that 'lag behind' the trends or that are innovative for their time – that more recent studies tend to focus on contemporary processes rather than past landscapes, are less descriptive and more analytical, and that theory and methods are more explicitly discussed or even become the central focus of inquiry. The means of relating form, sometimes inferring past processes, to analyses of process, often present-day processes, are methodologically very different, even though an outsider might not perceive such differences or their significances. This shift is arguably more about the shift away from an older 'denudation chronology', tracing the history of regional physical landscapes to a more analytical approach to glaciers and glaciation. In Glasgow, the 'alliance' between glacial geomorphology and topographic science and surveying techniques (Evans, 2009: 286), exemplified by Partington's dissertation research undertaken in Iceland, led to a strong connection between local, field-based details, and alertness to wider temporal and spatial contexts. This shift by no means indicates more 'quality' in the dissertations, because, even though the earlier dissertations are sometimes very descriptive, they still demonstrate high-level understanding of related elements in the landscape, which is maybe less specialised but arguably leads to a more holistic presentation of a given landscape as a whole.

Comparisons of different eras of coastal and fluvial geography research by undergraduate students also indicate a strong movement from regional landscape studies to landscape evolution studies through to more experimental projects, emphasising the modelling of landscapes, and making use of computational skills and technologies. That said, the significant role of field-based research persisted throughout different decades. In *A study of the evolution of coastal landforms at the mouth of the River Spey* (Robertson, 1978), the student used old manuscript maps, Ordnance Survey maps, and existing aerial photographs to trace the development of the 'the river-mouth landforms over the centuries' (Robertson, 1978: 11). The student combined the analysis of secondary sources with his own measurements of sediment 'drift' to evaluate how the beach was changing by injecting samples on the mid-tideline and taking wind speed, wind direction, wave height, and wave frequency data into consideration (Robertson, 1978: 37). He concluded:

> The principal finding was that the anticipated cyclic evolution of the landforms at the river-mouth was proved. The dominant drift direction has been shown to be westward. This means that, as the Tugnet spit

grows, the Kingston spit is gradually cut back and pushed landward. In the longer term, the whole form of the river-mouth has been altered as the seaward end of the delta has become choked with gravel and shingle leading to accelerated spit growth.

(Robertson, 1978: 62)

Robertson illustrated the mixed-method approaches used in many physical dissertations as it appeared early in such studies, as well as emphasising the central concern for the 'evolution of landforms'. Robertson's work was hence to some extent an example of a subdiscipline in transition: his methods still addressed the history of forms in the landscape and their evolution, but he was also already turning to contemporary processes and their measurement. Comparing this dissertation to Donna Baff's on *The effect of sediment sorting on the development of bedforms: A flume study using medium sands* (1994), the emphasis on processes becomes even more evident in the latter:

Both the present flume study and others with which it has been compared have shown that as flow intensity is increased a sequence of bedforms results. This sequence of bedforms and their characteristics are however, dependent on various factors of the flow and the sediment. As has been highlighted here sediment sorting is one such bed material factor that has important implications for bed configuration in sand bedded channels. The present flume experiment revealed that it is difficult to predict the bedform resulting from a particular set of flow conditions unless the bed material characteristics are known.

(Baff, 1994: 46)

Alongside the shift of some geomorphology projects becoming more experimental – using flume simulations – instead of descriptive, there was a simultaneous move towards more projects on the modelling of landscapes as well as towards computational and quantitative analyses. This move is evident from the research aims and objectives of more contemporary dissertations, such as Paul O'Connor's *Energy dissipation: A factor affecting physical erosion in free falling jet streams* (2014), where the scientific rigour stands in marked contrast to the descriptive evocations of earlier dissertations on water-eroded landscapes:

The overall aim of this study is to establish the relationship between height (H) of falling jet stream and unit flow (q) in knickpoint zones. These properties are evaluated with intention to reveal their ability to control landscape evolution and act as energy dissipaters in certain circumstances.

(O'Connor, 2014: 2)

Such changes are not only the consequence of better facilities and new technological and digital developments available to students but are often directly connected to specific staff members. Mentions of supervisors and other staff members in the acknowledgements indicate this link, but it is also striking to see the effect of appointments followed by projects of the newest lecturer or professor's specialisation within two years. For geomorphology, this is, for instance, recognisable with Trevor Hoey's appointment and, in O'Connor's case, the influence of Paul Bishop, an expert on what knickpoint migration reveals about landscape evolution, is maybe important. Yet again, the cause behind hiring specific members of staff must of course be partly prompted by current staff members recognising a lack of expertise on a certain subject, or, more positively framed, the specific value of recruiting someone with a new expertise never previously covered in Glasgow. Thus, specific appointments are related to disciplinary trends, and are often the turning point in the period of delay between trends in the broad, international academic community (and literature) and changes in the local undergraduate curriculum – and the dissertations stemming from it – of Geography as taught in Glasgow.

Besides disciplinary trends and the role of individual researchers, the confidence and previous experiences of students play a significant role in what they research and *how* they research their chosen subject matter. It is difficult, if not impossible, to pinpoint what comes first: a changing student population that prefers certain kinds of research practices or a changing Geography curriculum suddenly projecting a different message at outreach and recruitment events, subsequently attracting students 'matching' the revised curriculum. Trevor Hoey, former professor and Head of School at Glasgow, connects some recent changes in undergraduate dissertations to a decrease in previous outdoor experience of students:

> Certainly, [in school] Geography they were used to getting slightly more fieldwork than they do now; also in schools, I think time pressure at schools made it that less fieldwork is done. We did have students who had a little bit more field experience. When [Honours] classes were smaller, a higher proportion of the class would be sort of outdoor people who would go hillwalking or mountaineering in the weekends and so fieldwork was what they were easy with, they had all the gear, they knew what to do. Now we get students that are not so experienced in that sort of thing, and fieldwork doesn't matter quite that much to them. They don't see it as a particular reason to study Geography or Geomatics.
>
> (T. Hoey, interview, 2019)

Here, Hoey presents these changes in the previous experiences of students to some decrease – relatively recently – in the number of fieldwork-based dissertations, but he also sees changing complexity of the research practices and networks of physical geography as being responsible. Compared with professional physical geographers doing relatively small-scale, case study–based research circa 25 years ago, Hoey argues, there has now been a shift to multi-author teams undertaking big projects with very advanced and expensive equipment. This makes it more difficult to direct students to what is appropriate for an undergraduate dissertation:

> When I started you could show literature that had been published and say that this is literature on the same topic and even as an undergraduate they could do a scaled-down version of it. Because most of the literature 25 years ago was case-study based, relatively small-scale. ... So actually, in a way it has become harder to help students frame appropriate questions that are doing what we are saying in the assessment criteria, being critically informed, contemporary, and up-to-date because in order to do those sort of things – you can't as an undergraduate. It is really hard for them to take something from the literature and be like "oh I can see if I take this idea and this idea, I can actually solve this little problem". We [staff] can do that, but it is really difficult for students to do that ... I think the dissertation as it is structured, for someone who sees themselves as a physical scientist, does not quite do what they need.
>
> (T. Hoey, interview, 2019)

The professionalisation of physical geography and its research capacity perhaps has a discouraging effect for students instead of an *encouraging* effect, precisely because the 'smallness' in scale, time, finances, and human-power make the undergraduate dissertation less like an introduction to the *real thing* that it was previously in geomorphology (or indeed in some other disciplines and subdisciplines). It is a very striking discussion that raises questions about what exactly should be the aim of contemporary undergraduate dissertations, and how individualised a project it really should be.

The explosion of social and cultural geography from the mid-1990s

The subdiscipline with the greatest number of dissertations in the undergraduate dissertation archive in Glasgow has been identified as social geography. Accepting the complexities and limitations of categorisation, this subdiscipline has been taken to cover substantive foci of inquiry, such as migration, health, social cohesion, social media, and many, many more, but where there is a clear interest in social dimensions – social groups, uses, and spaces – bound up in these subject-matters. As Figure 4.5 demonstrates, there are two peaks perceptible when studying the number of social geography dissertations since the 1950s: one peak from the mid-1970s to the late-1980s, and a second, even stronger, peak from the late 1990s into the early years of the 2000s. This second peak, will be discussed in relation to the almost simultaneous growth in the number of *cultural* geography dissertations; but to understand what this subdisciplinary rise actually means, it is important briefly to address this earlier increase of social geography circa the mid-1970s to the late-1980s.

76 Becoming a geographer

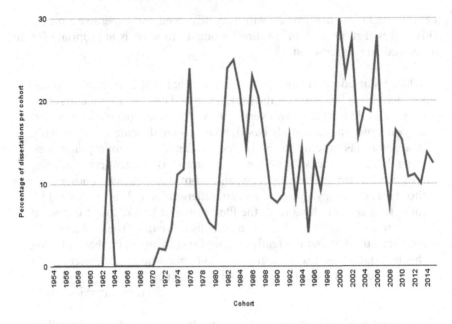

Figure 4.5 Number of social geography dissertations.

The first rise of social geography is inextricably connected to many movements of the late1960s and early-1970s which impacted academic geography, announcing the importance of addressing severe social issues:

> … what was at stake was also a liberal impulse towards social welfare and, for some, social activism. A wide range of research topics came under scrutiny, beneath the initial rubric of geographies of social problems (Herbert and Smith, 1989). … Other significant research topics include poverty and deprivation, social polarization, social exclusion, education, and housing and, in the consumer age of neoliberalism, geographies of leisure, tourism, sport and consumption.
>
> (Ley, 2009: 693)

As this quotation from *The Dictionary of Human Geography* (2009) explains, these social movements also influenced how geographers thought about their own role and about the role of academia as a whole. By taking the 'geographies of social problems' as a key focus, geographers could demonstrate the value of geography as an academic discipline, a significant benefit for a discipline struggling with its identity and status in academia, although clear tensions existed between those inclined to 'revolution' (Harvey, 1972) and those simply wishing the discipline to be more 'relevant' (Berry, 1972). Such a distinction can perhaps be detected in a few of the Glasgow dissertations. The social geography dissertations in the archive of the late 1970s and 1980s, as well as the urban geography dissertations (which also peaked during the 1980s), focus to some extent on those big social issues such as poverty and deprivation, crime, segregation, and housing, yet these offerings also share the stage with projects emphasising consumerism as well as recreation and tourism in the late-1970s and during the 1980s. The field of social geography was seemingly already well developed and positioned in the overall discipline and in the Glasgow undergraduate curriculum, based on the prevalence of all kinds of social geography dissertations. The foci of inquiry are diverse, although the type of research was often very similar: students were looking for patterns – social distribution patterns, retailing patterns, and mobility patterns – using methods appropriate for finding these patterns, emphasising quantitative conclusions and remarks. The 'dip' of social geography in the early-1990s is significant yet does not indicate a disappearance of social geography, with over 10% of the dissertations that can still be categorised as such. The increase of social geography dissertations from the late 1990s, however, represents nothing less than a landslide, both in numbers and the actual research endeavours of these many undergraduate students. The cohort of 2,000 presents an absolute peak of more than 30% of the students of this large cohort doing a social geography dissertation: the years after present high numbers as well, although slightly lower than 2000. To understand this popularity of social geography around the turn of the century, it is important to consider an almost simultaneous development in human geography: the rise of cultural geography. The peak of cultural geography in the dissertation archive lies just a few years later than this peak of social geography and appears almost 'out of nothing': some cultural geography dissertations are written in the late 1990s, followed by an explosion in the 2000s.

Addressing this simultaneous second rise of social geography and first rise of cultural geography in the undergraduate dissertations asks for some broader disciplinary reflections: going back to 1980. In this year, Peter Jackson published his paper 'A plea for cultural geography' (1980) in *Area*, explaining how British geographers were only just becoming aware of 'its potentialities' (Jackson, 1980: 110). Even within the Anglophone community, there seems to be a big discrepancy between national disciplinary traditions, and Jackson asks for a 'pact' between social and cultural geography in Britain:

> Cultural geography can finally only be of interest to the British geographical profession if it can successfully accomplish a rapprochement with social geography, in a joint commitment to study the spatial aspects of social organization and human culture – not just those aspects which are directly observable in the landscape.
>
> (Jackson, 1980: 113)

Jackson's paper was an important provocation for the direction of travel for work on social geography in Britain, leading to a significant reorientation and renaming of the relevant Institute of British Geographers study group (Philo, 1991).

Indeed, the boundaries of cultural geography and social geography turn out to be difficult to distinguish (Ley, 2009: 693). There are some examples of research on the far end of the spectrum that are obviously cultural; for instance, in dissertations such as *Dancing identities: The cultural geography of Highland dance* (Brogan, 2006), or evidently social, such as *Social cohesion in Kinning Park: To what extent does a community exist?* (Clark, 2006). However, with the 'culturalisation' of many branches of geography (Gregory et al., 2009: 129), social geography has not become far away from 'culture' and cultural analysis. This section does not attempt to discuss the rise of social *and* cultural geography in general, however, but aims to add two things to the narratives about this rise. First of all, there are many research projects written about in undergraduate dissertations that are read by almost no one, yet are, in their quality and innovativeness, impressive geographical knowledge productions – and nowhere perhaps is this more obvious than in a number of the social-cultural geography dissertations from the mid-1990s and early 2000s. Second, it is important to emphasise the positive influence of individual supervisors, but also of small changes made in the undergraduate curriculum that were an impetus for students to research certain themes. Analysis of the undergraduate dissertations demonstrates the importance of teaching practices in universities, opening a window on how different accents in the courses, different Honours options, or different members of staff all help students to discover different, new interests or to realise the opportunities to research certain themes that they otherwise might not have considered 'geographical enough'. The near-simultaneous arrival in Glasgow of Chris Philo, Paul Routledge, and Jo Sharp marked a new impetus for both research and teaching in the department, invigorating and in various ways melding social and cultural geography foci in the undergraduate dissertations. Sharp ran an Honours course called 'Cultural Geography', Routledge taught courses in which 'social movements' were centralised, and Philo commenced an Honours course called 'The Social Geography of "Outsiders"' from the academic year 1997 1998.

This second peak of social geography dissertations focuses largely on the latter, societal 'outsiders', although the breadth of what might be considered as 'outsiders' still makes this collection of dissertations highly diverse: from elderly people, children and women, via homeless people, differently sized people, people with specific health conditions, through even animal geographies. Some of these 'outsiders' perspectives were already researched earlier, for instance, by Peter Soward in his dissertation *An investigation into the cognitive maps of the City of Glasgow as held by disabled persons* (1994):

> Due to the nature of this study it is important to remain detached from the sensitive issues found in conjunction with disability thus remaining unbiased. This results in a fairly cold, analytical treatment of an area

that effects real people. This is unfortunately necessary to achieve results of value.

(Soward, 1994: 2)

This dissertation is *about* outsiders, but Soward explicitly stated that it had to be written in a 'detached' way to have 'value'. During the later wave of social geography dissertations, many students justified an opposed approach, deliberately aiming for involvement, a more 'intimate' attitude of the student-researcher. A fellow 1994 student, Aileen Donaldson, also wrote her dissertation, entitled *Reasons why the retired elderly migrate to Crieff* (1994), about a specific social group often seen as 'outsiders' in geographical research, but her approach was not a social geography one but more connected to the subfield of population geography:

> Geographers have much to contribute to this field of study, conducting research in areas such as demographic trends, migratory patterns or settlement structures of the elderly. ... In order to investigate one specific component of population geography, it is necessary to locate research concerning the elderly within geography as a whole.
>
> (Donaldson, 1994: 2)

Donaldson also reflected on the position of 'the elderly' as a focus of inquiry in geography:

> Geography is a multi-faceted social science, therefore gerontological issues tackled from a geographer's perspective highlight both public policy and theoretical implications.
>
> (Donaldson, 1994: 3)

She rightfully addressed how 'the elderly' as a social group will prompt attention from geographers in years to come:

> Looking ahead to the twenty-first century, early retirement, and government health legislation will alter the patterns of migration both positively and negatively. There is no doubt that for geographers and society in general to attain a deeper understanding of the elderly's behavior patterns, more age specific research on a micro scale, such as this study of Crieff is required. ... As the number of pensioners continues to rise, research into the elderly must be a priority for society and government alike to ensure maximum quality of life for our elderly and efficient use of national resources.
>
> (Donaldson, 1994: 30)

These observations from Donaldson are in themselves instructive, revealing a student with keen (sub)disciplinary knowledge and able to locate her own study accordingly, but it is telling to contrast these contributions by Soward and Donaldson with the kinds of social geography treatment lent to 'marginal' groupings a few years later. Four years on, for instance, Katrina Slater also addressed the elderly in her dissertation *Geographies of the elderly in residential care: Closing geographical lifespace?* (1998), but now wearing a very obvious social geography 'hat' alert to the broader span of what might be considered under this subdisciplinary heading. Indeed, she compared her 'target audience' to other, more 'popular' groups of outsiders:

> The study of old age in academia has been going on now for quite some time, concentrating on the physiological, psychological, social, economic and demographic aspects of ageing. So far this has been relatively neglected within the field of geography as many see this as not 'exciting' enough unlike the 'in' subjects such as 'children's geographies' or 'gay and lesbian geographies'.
> (Slater, 1998: 1)

In line with Slater's last remark, children get a more obvious place on the 'geographical stage' in these years, certainly according to the evidence of the undergraduate dissertation archive, than do the elderly. One example is *'The terrible twos': A social geography of childcare* (Gallacher, 2002).[4] In her research, Gallacher analysed the daily geographies of childcare:

> As such, they [young children] cannot be ignored by Geography, written off as 'pre-social' and inconsequential. The internal, psychological space of individuals is also intimately bound up in the social relations in Stepping Stones [a nursery], informing the manifold of styles of control and resistance that continually contest and reconstitute social space.
> (Gallacher, 2002: 45)

Gallacher also explained how young children are underrepresented, and even more poignantly, underestimated in geographical research:

> The existing geographical literature seriously underestimates young children: the under-fives should be seen as 'real' social agents whose actions do have relevance outside the family. My focus here is on toddlers because they are doubly stripped of social agency.
> (Gallacher, 2002: 8)

This last quotation demonstrates how the social geography of outsiders as revealed in the dissertations was often strongly connected to social theory – with its recurrent emphasis on theorising the capacities and limits of human (social) agency – as well as to a sometimes-political language concerning agency, responsibility, and structure and power relationships. The point is that social geography dissertations studying 'outsiders' commonly operated at a different theoretical and conceptual level from their loosely 'social' predecessors: the focus was less about recognising and describing behaviour and other patterns, but rather about the philosophical, sociological, and geographical framings, causes, consequences, and implications of such patterns, often studied on the microscale.

Another example of a dissertation that addresses children is Victoria Smillie's dissertation entitled *The geographies of a dance school* (2010). Similar to Gallacher, Smillie asks questions concerning power over and agency of children in a particular space, but expressly connects these themes to the concept of embodiment:

> Through an awareness of space and an understanding of their physical body within that space, the children of the Dance School learn how to embody the environments which they inhabit. Dance allows them to understand the physical, flesh and bones working of their body as well as the way it can be used. ... In saying that, I do not feel that this project has become 'just' another version of existent children's geographies studies of schools, after-school clubs and play centres. It certainly echoes some of the issues raised by such studies, [but] the dance angle does seriously inflect these studies by foregrounding questions about the bodies of children, and their ability to embody, perform and resist.
>
> (Smillie, 2010: 35–36)

Smillie's innovative research here demonstrates how the 'microscale' of the body and the related concept of embodiment have recently acquired a place in academic geography. The latter is a concept that bridges social geography and cultural geography, and it is strongly embedded in conceptual approaches such as non-representational theory, focusing on practice and performance (Anderson, 2009: 505). The rise of cultural geography is often distinguished across two strands, with the first one drawing on work tackling questions concerning the sociology of culture and specifically researching cultures and subcultures in modern life, where the second strand is seen as 'cultural' both by topic and method (Crang, 2009: 131), placing geography in a more humanities-like tradition, akin to how other geographers connect their discipline to either the natural sciences or the social sciences. This second strand is recognisable in the early years of the century, with dissertations on, for instance, dance, such as Smillie's, or music. Isla Forsyth's dissertation *The place of folk music in cultural geography: Dick Gaughan, a modern day Scottish prophet* (2006) explicitly discusses what could be the role of music in the discipline:

> From the information gathered by the research and the reading carried out beforehand, it appears that folk music can have a potentially important role within the discipline of Geography. It offers the chance to experience and awaken a more dynamic discipline that waves goodbye to the elitist static notion of cultural geography so far dominating work, which

is not representative of how people experience their culture. ... This can be used in bringing a new and more holistic geographic understanding of how people interpret their culture and environment, which in turns contributes to their wider sense of identity.

(Forsyth, 2006: 22)

Cultural geography, and specifically the newly emerging field of geohumanities, does not ignore the physical landscape, and even sometimes goes back to the 'classic' idea of geography as a discipline studying the relationship between humans and their physical environment. This relationship is, however, then researched in a very different way from what was preferred by their regional geography predecessors. For instance, in the dissertation *The Mississippi Delta, its landscapes and the Blues* (Astill, 2010), the student-researcher connects social-historical research, the analysis of the physical landscape, and cultural analysis of Blues music in a compelling way:

If we are to explain the extent to which the physical geography of the Delta influences Blues music, it is clear that there is a great extent of music designed around the concept of flooding which has been discussed throughout this research. ... The Delta itself provided an ideal breeding ground for blues fundamentally based on racial inequality and discrimination. The social climate is reproduced in associated soundscapes through the music, lyrics referencing inequalities were found to be relatively common, the social background of the Delta thus proves a significant influence to the Delta Blues[;] after all the Blues stemmed from suppression faced by slaves.

(Astill, 2010: n.p.)

These examples of very obviously culturally turned research projects demonstrate that geography as a whole has not necessarily strayed *so* far away from certain root conceptualisations and justifications about what geography 'should be' from decades ago. However, the foci of inquiry, methodology, and conceptual frameworks are in many cases very different. With its more recent allegiance to the humanities, the research done under the banner of geography might be even more diverse than ever before, with inevitable knock-ons for the diversity of subject-matters researched by recent undergraduate dissertations, even when positioned under the subdisciplinary umbrellas of social and cultural geography.

Conclusion

The analysis of the research endeavours undertaken in more than 2,600 locally produced undergraduate dissertations – turning to their substantive subject-matters, allied to questions about the relations between physical and human geography, and about subdisciplinary allegiances – has provided an opportunity to add the experiences of almost as many geographers-in-the-making to existing narratives of 'the' history of geography. It is interesting to remark that many (sub)disciplinary trends that are generally recognised in the history of Anglophone geography are often visible in the dissertation collection, but with a substantial delay. What is argued here is that such delays might be influenced by the fact that a trend only becomes recognisable in undergraduate research when academic members are appointed who are part of, or 'early adapters' of, specific (sub)disciplinary changes and trends. Their arrival at the university is an opportunity to bring new accents, new 'worlds', to the undergraduate curriculum. This does not mean that it is only such 'newcomers' who play a significant role in such innovative endeavours, however, as such a claim would deny that 'ongoing' staff cannot engage with, embrace, and encourage work reflecting new (sub)disciplinary developments. It is also plausible that hiring committees, including current academic members of staff, explicitly recognise a lack of expertise in a specific subdiscipline or an opportunity to appoint new staff *au fait* with newest trends.

The undergraduate curriculum is a place where the subject of geography is taught as a unity, with human geography and physical geography (usually) 'peacefully co-existing' in this educational framework. However, there are convergent and divergent movements perceptible through time, with specific foci of inquiry providing bridges at times and epistemological and ontological conceptions driving them apart at others. When studying a longer period of time, there are some clear breaking points recognisable: turning points perhaps for the research that undergraduate students are doing – or the ways in which they are justifying, explaining, and reflecting on their research activities. For the undergraduate dissertation archive, one decisive point is the regional dissertation requirement ending in the 1970s. After this time, more dissertations include hypotheses, explicit research aims and questions, and some explanation of the connections between such objectives and the methods used. The second major breaking point is during the mid-1990s: parallel but independent developments in the department caused renewed research opportunities (for instance, because of improvements of the research facilities and equipment for physical geographers) and renewed attention for geographical theory, methodology, and reflective–reflexive practices.

Studying the geographical knowledge that has been produced by undergraduate students provides a view into the hybrid space between formal academic geography, taught by employed geographers, researching and teaching within the context of the discipline, and the non-specialist – or, rather, the as-yet non-expert or, better perhaps, becoming-expert – geography of students. This discussion creates a sometimes predictable, but other times intriguing and surprising perspective. The undergraduate dissertation is a relatively 'free form', and as such highly distinctive from almost all other work that students do, such as coursework assessments, examinations, and lab practicals. Within a limited time period with limited means, students have to inject much of 'themselves' in their work – sometimes by choosing a focus of inquiry close to their hearts and other times in the experience of independently executing all different aspects of the research.

Intermezzo 4: Social Justice

From the 1970s, dissertations appear that display some kind of 'social concern' emphasising problems such as poverty and housing. The expanding scope in social geography, discussed earlier, goes hand-in-hand with the rise of radical and Marxist conceptual frameworks. The following two dissertations written by students from the 2010 cohort are not only exemplary of a 'trend' in the language of a dissertation title (a short quotation, followed by a colon and a more general description of the focus of inquiry) but also of a new way of emphasising the concept of social justice. Louise Boyle's dissertation, entitled *'Dear green place': Space, place and communication within the 'Towards Transition Glasgow' Network* (2010), explored a specific network, based on archival material, spatial mapping, questionnaires, interviews, and participant observation. This network analysis not only offered insights into the involvement and ambitions of members of the network but also aimed to demonstrate structural power relations and social inequalities:

> Glasgow represents not only a facet of motivation through which people are given strength to pursue their interests and ambitions but a unique platform as a convergence space facilitated by constant pockets of activity; extensive connections; common platforms of collective action; overlapping circuits of solidarity and sites of contestation in terms of power and social relations relating to knowledge, dominance and social stature.
> (Boyle, 2010: 33)

The references here to social movements, 'convergence space' and 'solidarities' signal the influence of Paul Routledge, mentioned earlier, and also an emerging Glasgow speciality with respect to what has more recently been termed 'spatial politics'. Joanne Armour, also a student from 2010 who wrote her dissertation *"Think global, eat local": Insights to food localization in East Ayrshire* in 2010, undertook a very different kind of project, emphasising food localisation and connecting this topic to social justice:

> One of the fundamental concerns of the social justice literature is that localization will exacerbate the already unfair and detrimental social problems that will be in place in certain localities.
> (Armour, 2010: 30)

Based on a large number of questionnaires and interviews, she concluded that social justice issues are often overlooked at the household and regional levels

(Armour, 2010: 36). An overview of quotations from participants demonstrated a mixed opinion about the success of local food networks, with some, for instance, emphasising that only a small group within the community 'profits' from it. These two examples demonstrate that research questions can combine social research with more philosophical and conceptual analyses concerning social justice. It is also exemplary of research that is ethical in nature: demonstrating ethics not as a 'side issue' of doing research, but as a very central concept driving the research.

Notes

1 Technically, 'regional geography' is *not* a subdiscipline given how it was understood as where the 'systematic geographies' come together to reveal the characteristics and, potentially, the unity of a whole region. However, the regional geography dissertations are so distinctive from the others that I also used them as a subgroup in this analysis of subdisciplines.
2 For instance, *Geomorphology of area north and east of Helensburgh* (Robertson, 1965) and *Geomorphology of Strathallan* (Smith, 1967).
3 The relationship between Glasgow Geography and Glasgow Geology is itself complex and tangled (Leake and Bishop, 2009). From 2004 what had been the Division of Earth Sciences (previously Geology) was formally brought into the Department of Geography and Geomatics (previously Topographic Science) to create a unified Department (now School) of Geographical and Earth Sciences.
4 This dissertation was eventually reworked for publication in the journal *Children's Geographies* (Gallacher, 2005).

Bibliography

Adamson, A., 1998. *"The view from their window": How neighbourhoods affect attitudes to education and outlook on life*. Undergraduate Dissertation, University of Glasgow.
Anderson, B., 2009. 'Non-representational Theory'. In: Gregory, D., Johnston, R., Pratt, G., Watts, M., Whatmore, S., (eds). *The Dictionary of Human Geography* (5th ed.). Hoboken: Wiley-Blackwell, 503–505.
Armour, J., 2010. *"Think global, eat local": Insights to food localization in East Ayrshire*. Undergraduate Dissertation, University of Glasgow.
Astill, R., 2010. *The Mississippi Delta, its landscapes and the Blues*. Undergraduate Dissertation, University of Glasgow.
Baff, D.G., 1994. *The effect of sediment sorting on the development of bedforms: A flume study using medium sands*. Undergraduate Dissertation, University of Glasgow.
Berry, B.J.L., 1972. '"Revolutionary and counter revolutionary theory in geography" – a ghetto commentary'. *Antipode*, 4(2), 31–32.
Boyle, L., 2010. *"Dear green place": Space, place and communication within the 'Towards Transition Glasgow' Network*. Undergraduate Dissertation, University of Glasgow.
Brogan, K., 2006. *Dancing identities: The cultural geography of Highland dance*. Undergraduate Dissertation, University of Glasgow.
Clark, L., 2006. *Social cohesion in Kinning Park: To what extent does a community exist?* Undergraduate Dissertation, University of Glasgow.
Donaldson, A.M., 1994. *Reasons why the retired elderly migrate to Crieff*. Undergraduate Dissertation, University of Glasgow.

Evans, D.J.A., 2009. 'Glacial geomorphology at Glasgow'. *Scottish Geographical Journal*, 125(3-4), 285-320.
Forsyth, I., 2006. *The place of folk music in cultural geography: Dick Gaughan, a modern day Scottish prophet.* Undergraduate Dissertation, University of Glasgow.
Gallacher, L.A., 2002. *"The Terrible Two's": A social geography of childcare.* Undergraduate Dissertation, University of Glasgow.
Gallacher, L.A., 2005. 'The terrible twos': gaining control in the nursery? *Children's Geographies*, 3(2), 243-264.
Gregory, D., Johnston, R., Pratt, G., Watts, M., Whatmore, S. (eds) 2009. *The Dictionary of Human Geography*. (5th ed.). Hoboken: Wiley-Blackwell.
Jackson, P., 1980. 'A Plea for Cultural Geography'. *Area*, 12(2), 110-113.
Ley, D., 2009. 'Social Geography'. In: Gregory, D., Johnston, R., Pratt, G., Watts, M., Whatmore, S. (eds). *The Dictionary of Human Geography*. (5th ed.) Hoboken: Wiley-Blackwell, 692-693.
McAuley, J.B., 1982. *The Impact of man on the water quality within the South Calder Basin.* Undergraduate Dissertation, University of Glasgow.
O'Connor, P., 2014. *Energy dissipation: A factor affecting physical erosion in free falling jet streams.* Undergraduate Dissertation, University of Glasgow
Partington, E.A., 1994. *The comparison and analysis of two methods of collating ablation data for the snout of Sandfellsjokull, Myrdalsjokull Ice-cap, southern Iceland.* Undergraduate Dissertation, University of Glasgow.
Philo, C., 1991. *New Words, New Worlds: Reconceptualising social and cultural geography: Proceedings of a conference.* Lampeter: St David's University College.
Philo, C., Ernste, H., 2009. 'Determinism/environmental determinism'. In: Thrift, N.J. and Kitchin, R. (eds). *International Encyclopedia of Human Geography*. Amsterdam: Elsevier, 102-110.
Rankin, J.B., 1962. *Morphology of the Ardoch Estate area with special reference to drift morphology and raised beaches.* Undergraduate Dissertation, University of Glasgow.
Robertson, A.M., 1978. *A study of the evolution of coastal landforms at the mouth of the River Spey.* Undergraduate Dissertation, University of Glasgow.
Scott, L., 2010. *The role and perception of environmental journalistic cartography: Eyjafjallajokull volcanic eruption case study.* Undergraduate Dissertation, University of Glasgow.
Sinclair, M., 1978. *The glacial geomorphology of the middle Endrick Valley - West central Scotland.* Undergraduate Dissertation, University of Glasgow.
Slater, K.A., 1998. *Geographies of the elderly in residential care: Closing geographical lifespace?* Undergraduate Dissertation, University of Glasgow.
Smillie, V., 2010. *The geographies of a dance school.* Undergraduate Dissertation, University of Glasgow.
Smith, B., 1998. *Geographic dimensions of space and place involved in social relationships between 'incomers' and 'locals' within a small Highland community.* Undergraduate Dissertation, University of Glasgow.
Soward, P., 1994. *An investigation into the cognitive maps of the City of Glasgow as held by disabled persons.* Undergraduate Dissertation, University of Glasgow.

5 Geographical traditions versus innovations

Students as drivers of disciplinary change

The question of what geography actually *is*, or *should be*, is possibly a more relevant question for geographers-in-the-making, compared to students of other academic disciplines, since debate on the very 'nature' of academic geography is a constant factor in the discipline's past. There are, however, important shifts over time as well as individual differences in how implicitly or explicitly such questions are discussed in the dissertation, in how 'standardised' are reflections on the discipline and one's own work, and in the attention paid to situating research within a conceptual and theoretical framework justifying and explaining the epistemological and moral value of the project. The reflective notes in dissertations on what geography 'should be' and what a geography student 'should do' are telling of the complexity and hybridity of academic geography, but also demonstrate how one department and its staff translates its own status, ambitions, and readiness into new ideas and new methods for teaching and supervision. Changes in this awareness of the discipline and of 'oneself' as a geography student might also be connected to a changing student population. This chapter explores these different levels of disciplinary awareness, combined with an inquiry into the conceptual frameworks adopted, more or less explicitly, in the University of Glasgow's collection of geography undergraduate dissertations. This leads to a longitudinal perspective on how apprentice geographers 'find their ways' into their discipline, but also how in the process they may change, reshape, and challenge the contemporary 'versions' of geography in which that they are being educated.

Disciplinary awareness

> Geography has often been defined as the interaction of man [*sic*] and his environment – which is broadly the essence of the matter. In geography we do not study man alone – a single entity – but rather a community of men, all inter-related and complementary in their functions. Environment should be interpreted in its widest sense to include not only the purely physical like climate and vegetation, but also such factors as communications, orientation towards and relations with other communities. The relationship between the community and its region is then the fundamental study in geography. It is itself an organism, two different aspects

of which are represented by community and region. This organism is ever changing and growing where sufficient stimulus is provided and where we find such a healthy development we may say that there occurs a true natural geographical region. In all parts of the world we find such natural regions, and the most flourishing are, not isolated from all others, but in constant contact with them.

(Munro, 1954: 1)

This quotation from Munro's dissertation *Aviation as a new factor in geography* (1954) about geography as a discipline is unique for the dissertations written in the 1950s and 1960s, since such meta-level reflections on the nature of geography – or indeed of its subdisciplinary fields – are extremely scarce in these earlier years of the Glasgow dissertation. In hindsight, it is often possible to place dissertations in a specific disciplinary tradition, drawing as they do (if not all that self-consciously) upon specific bodies of concepts and methods, but students in these early decades rarely explained how their work related to the wider discipline, nor why their research might be relevant within (or beyond) the academy. Standard disciplinary histories often provide narratives about shifts in methodologies and concepts used, sometimes conceived in terms of paradigms or relatively coherent 'isms' and 'ologies' (e.g. Cresswell, 2013 and Johnston and Sidaway, 2016). Unsurprisingly, the dissertations across the years roughly mirror these frequently described 'trend lines': the academic staff of a department shape and mould the undergraduate curriculum – sometimes more conservatively, sometimes more innovatively, depending on the particular staff members, their own generation and background – which is therefore not a separate reality divorced from wider academic trajectories. However, analysing levels of disciplinary awareness presented in the dissertations, asking about how students have seen the discipline (and its own subdivisions) of which they are part and hence the place of their own work within this discipline and its transforming conceptual landscape, opens a new window on how a discipline is indeed made, circulated, received, and perhaps – if here only in minor ways – recast.

Disciplinary awareness can be distinguished in thoughts about *what* geographers are studying and *how* they do this. The following two quotations demonstrate two possible, quite different versions:

In recent years the application of statistical techniques has become commonplace in helping to solve problems of a geographical nature. As this is a geographical problem being studied in this paper, statistics shall be used in accordance with this trend.

(Hastings, 1990: 27)

… mainstream Geography all but ignored children until the 1990s.

(Moore, 2006: 9)

The two examples reveal either fitting in with the supposed conventions of a discipline or seeking novelty. The second quotation, about children's geographies, makes a claim about what geography is *not* doing or including (enough), according to this student, with the implication that something *new* is indeed required: a new move in terms of what gets studied. The first quotation is a justification of the methodology used in a research project, saying that, by following an already established orthodoxy, although arguably by 1990 a statistical orientation was already a bit dated or at least itself heavily critiqued within urban geography, the geographical problem posed by the student can be solved. The contrast between 'following' the tradition and trying to change the discipline, or at least to fill a gap within the discipline, is probably down to differences in students' own attitudes and skills in tandem with the influence of their supervisors. It also might suggest an increased emphasis on 'novelty' in research, rather than following the 'rules', suggesting a really quite different sense of what a student-geographer might or should be able to contribute, itself reflective of how the overall dissertation task has been presented to them in preparatory classes and by their supervisors. Especially in the cohorts from the late 2000s and early 2010s, students explicitly use the language of 'originality' and 'novelty' in justifying all sorts of inquiries, probably linked to the rise of formal guidance that itself spoke about students needing to seek out 'gaps' in the literature that their dissertation might address. Such guidance inevitably demanded greater awareness of disciplinary and subdisciplinary literatures, including the capacity to encompass, describe, and critique such literatures.

Adding this longitudinal analysis of views expressed by geographers-in-the-making to disciplinary narratives shines a fresh light on two aspects of the discipline. First, it exposes the 'step changes' in what undergraduate geography students are taught occurring between a few cohorts of students, maybe only a few years apart, inevitably influencing subsequent generations of postgraduate students and early career academics; and second, the obvious change in language, openness, and awareness (of wider traditions, their changes and challenges) over time. The skills of reflecting on one's own role as researcher and about research methodology, tied in both respects to matters of research ethics, clearly took a giant leap around the turn of the century for these Glasgow undergraduates. Echoing changes in the cultural–intellectual conceptions (held by teaching staff) of what researchers 'should do', and in epistemological arguments around what is 'geographical knowledge', not only the questions asked and methods used by undergraduate students changed, so too did their reflexive engagement. Such changes from the early 2000s are perhaps even more perceptible in the attention given to a 'methodology chapter', the formation of not only a research question but also the stating of explicit 'research objectives' mapped across into methods, and also the including of sections about ethics and one's own positionality. This is not an entirely new insight, but there is arguably novelty in realising that not 'only' the vanguard of the discipline was starting to engage with such reflectivity and reflexivity, but so too were geography students, meaning that these novice geographers now started to approach their work with greater levels of both self-awareness and disciplinary awareness than was usually true of earlier students. It is not only the changing topics and concepts – for instance, the rise of dissertations on climate change and sense of place – that tell us what geography is, but also what geographical knowledge and, moreover, disciplinary awareness so many of these individual geographers – perhaps not continuing in academic careers, but geographers nonetheless – have come to display in their first big research project.

The disciplinary awareness of students is often expressed in brief remarks about the history of geography or that of a specific geographical subdiscipline, or indeed of the history of a particular body of geographical knowledge (maybe including people other than professional geographers). The time scale of these short historical analyses is diverse, from mentioning the ancient Greeks to summarising the most up-to-date disciplinary developments. Sometimes this history is told by means of referring to the 'grand names' in geography's history:

> W.M. Davis' slope profile development model was termed the 'Geographical Cycle'. ... The criticisms of the Davisian system have been widely expressed and include such points as denudation can only be important when a landmass is stable; and also that streams undergo 2 phases of activity namely rapid incision and then virtual dormance once its grade has been attained. Therfor [sic], his 1930 ideas are often overlooked but his name is still associated with the overall cycle of erosion and its flattening concept and development of a peneplain.
> (Mauritzen, 1994: 6–7)

Others refer to more 'amateurish' knowledge and interests:

> In the late 18th and 19th centuries many farmers and amateur geologists became interested in the reasons for the distribution of the different types of soil. They came to the general conclusion that there was a direct and fairly simple relationship between the character of the soil and the nature of the bedrock beneath it.
> (Watson, 1994: 3)

Although the scope of a standard chapter dedicated to the analysis of academic literature and the contextualisation of one's research increased, especially from the 1990s onwards, this *historical* analysis is often limited to a few sentences. Significantly more attention is paid to more contemporary contextualisation of the research question, both within academic geography and within the wider academic context.

Geographical traditions versus innovations

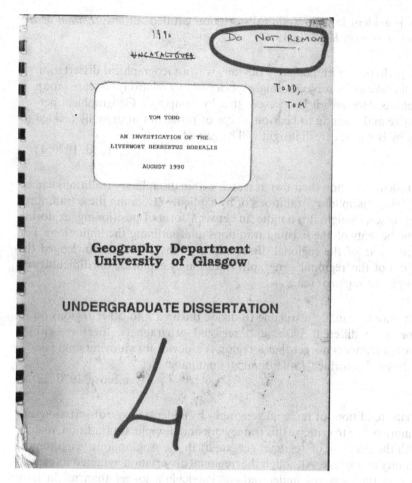

Figure 5.1 Dissertation cover (Todd, 1990).

To explain what they are doing in their research for their undergraduate dissertation, some students explicitly position themselves as *geographers* in a wider academic field. Especially in physical geography dissertations, students refer to geologists, as demonstrated in the quotation just mentioned. There are two other disciplines regularly mentioned besides geologists by geography students: psychology and botany. It may come as a surprise that these two disciplines were specifically mentioned several times by students situating their own research, whereas there are disciplines that seem more closely related to geography (for instance, sociology). However, it might be the case that, especially when students decide to choose an 'odd' dissertation topic, not obviously geographical in nature, they need to explain how and why this specific subject might be geographical as well.

This is evident in Tom Todd's dissertation entitled *An investigation of the liverwort Herbertus Borealis* (1990):

> It is perhaps not immediately obvious why a Geographical dissertation should take as its subject a single species from a relatively obscure group of plants. However it is believed, that by bringing a Geographical perspective and training to bear on a type of problem traditionally tackled only by Botanists, fresh insight will be gained.
>
> (Todd, 1990: 1)

How students position their own research within disciplinary traditions can be by justifying disciplinary traditions or by explicitly criticising these traditions. In many cases, though, it is a more 'in between' form of positioning, exploring both the benefits of the existing traditions and outlining the limitations. For instance, some of the regional dissertations of the 1970s acknowledged the fading out of the 'regional' geography tradition by exploring the difficulties of working at the regional scale:

> ... it may be equally justifiable to divide the area into other regions on some quite different basis, and regional geographers, together with regional planners, forget that a region is a device for studying man [*sic*] on the earth's surface, rather than an immutable fact.
>
> (Ambrose, 1970: 22)

Such 'late' renditions of regional geography by undergraduate dissertations are often more prone to criticise this framework and disciplinary tradition, matching with the decrease of regional geography in the Anglophone geographical community as a whole. Although the 'regional dissertation' retained its central position in the Glasgow undergraduate curriculum longer than might have been expected given claims in the published historiographies, it had yet to pass a few years of increasing critique and discussion of 'the region' as an organising concept before it became extinct.

The cohort of 1998: disciplinary awareness and conceptual framing

There is a notable increase of reflections on the position of students' own dissertation from the mid-1990s. In this section one cohort exemplary of this 'breaking point' – 1998 – will be discussed in further detail. From then, dissertations bear witness to disciplinary awareness and conceptual framing much more consistently and explicitly. By reflecting on the history of geography, or the history of a specific subdiscipline or concept, students indeed place themselves in a geographical timeline. Simultaneously, by framing their research in a specific tradition, they take a more or less self-aware epistemological and sometimes ethical 'position' within the discipline. Often such historical–philosophical justifications go hand-in-hand: by telling a specific (short) narrative about the history of geography, they address the need or 'overlookedness' of their own research.

The 69 dissertations written in 1998 explore different corners of the planet as well as corners of the geographical discipline. It is nonetheless obvious that there are some 'hot topics' addressed: for instance, crime and the fear of crime, the geographies of 'outsiders' and conservation. Whereas before students may have commented on the task of 'the geographer' or what 'geography has neglected', many students in the 1998 explicitly demonstrated some *subdisciplinary* awareness: they were not only reflecting on geography as a whole, but often more specifically reflecting on the history and philosophy of a specific subdiscipline of immediate relevance to their inquiry. This might indicate a stronger compartmentalisation within the discipline:

> The field of urban geography encapsulates some of the most complex philosophical theory and debates in the discipline of geography today.
> (Gaunt, 1998: 9)

> The study of deformation tills is an expanding field of glaciology. ... The study into deformation tills is an important area of study as it is linked to many parts of glacial & quaternary research.
> (Shiels, 1998: 1)

It is not only the way students referred to the past of their 'own' subdiscipline, but the fact that almost all of them reflect explicitly on 'their' subdisciplinary field of concern demonstrates a difference between pre-1990s and later dissertations. For human and physical geographers alike, a geography dissertation needed to include some reflections on what academic predecessors were doing. This is an awareness that fits into the Glasgow-specific curricular changes that happened during the mid- and late 1990s, ones that might also be shaped by renewed attention to the history of geographical knowledge in the wider academic community (e.g. Domosh, 1991; Livingstone, 1992). That said, whereas in the community of British geography some geographers now focus their research *on* the history and/or philosophy of geography itself, not a single undergraduate dissertation in the collection takes such a topic as their main research subject! This meta-analysis thus became a standard element of the geography undergraduate dissertation, yet did not convince any student to take it on as their actual focus of inquiry.

Besides the historical awareness of geography, students of the 1998 cohort paid attention to the conceptualisation of their research by addressing the theoretical framework they are wielding. Again, the students writing their dissertations in 1998 were one of the first cohorts who do this on such a widespread scale; likely influenced by the third-year core course 'Geographic Thought', run since the academic year 1996–1997. Many students include Marxist or postmodernist thought in their framework:

> The emergence of radical geography in the 1970's with its strong Marxist connotation, initiated a strong attack on the conservative bases of positivism. Whilst poverty, and disadvantage have been attributed as conditions from which delinquency emerges, radical theorists delve deeper and see the origins of poverty and disadvantage in the class-based mode of production. The city is considered to play an important role as it provides the conditions necessary for the perpetuation of the economic bases, which in Western society is the capitalist mode of production.
> (Black, 1998: 7)

> My dissertation has been largely influenced by the ideas articulated by the postmodern approach to geography. Firstly, I have avoided using a metanarrative or a totalising discourse to structure my work around. Instead I am investigating the different discourses surrounding my theme, becoming more open to other philosophy and perspectives.
> (Kitchingham, 1998: 11)

As demonstrated by these two quotations from 1998, often the conceptualisation for a project was embedded in a historical contextualisation of specific 'streams' or approaches. Some students, however, presented a very well-developed conceptual background to their own research: for instance, Katy Kitchingham did so in her dissertation entitled *Investigating 'nature' – A case study of Crawford Lake Conservation Area in Ontario, Canada* (1998).

Kitchingham remarked upon the history of the concept of landscape by referring to the ideas of Thoreau and Muir, yet also introduced social constructionism and feminist thought about landscapes:

> Landscapes are often seen in terms of the female body and the beauty of nature, by the dominant gaze of white, masculine, heterosexual knowledge. There are subtle indications of this at Crawford Lake. The terminology used to describe the different elements ranges from *beautiful, lush* and *pristine* on the feminine side to *rugged* and *majestic* on the masculine.
> (Kitchingham, 1998: 31)

She very clearly described what she was doing in her research, as well as what she was explicitly *not* doing:

> Looking at landscape from a cultural, rather than a physical geography perspective, it is clear that postmodernism has had a profound influence on how landscape is perceived.... Landscapes are seen as constructions, representations of a material environment that have usually been created by the dominant discourse in society.
> (Kitchingham, 1998: 6)

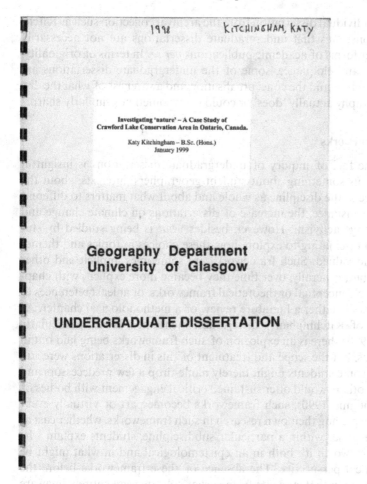

Figure 5.2 Dissertation cover (Kitchingham, 1998).

Kitchingham connected subdisciplinary perspectives (landscape geographies in effect) to conceptual frameworks, but was also not afraid to disagree with dominant trends or attitudes, and intriguingly she anticipated how social constructionist approaches have since been critiqued and reshaped by new 'materialist' orientations:

> The philosophical beliefs that underpin my dissertation can be described as both antinaturalist and realist.... Although I embrace the new ways of thinking about human society that have been brought to the discipline by, for example, postmodernism and feminism, I still firmly believe that there is a material side to the Earth's existence that is not entirely created in our minds and systems of communications.
>
> (Kitchingham, 1998: 10)

Exploring an individual dissertation from the archival collection such as Kitchingham's demonstrates that undergraduate dissertations are not necessarily inferior to other forms of academic publications *per se*. In terms of originality, argumentation, and eloquence, some of the undergraduate dissertations are surprisingly effective, and the conceptualisation and awareness of what the discipline of geography actually 'does' or could do is sometimes similarly sharp.

Conceptual frameworks

Researching the foci of inquiry of undergraduate dissertations is insightful because it tells us something about student-geographers' interests, about the gaps they notice in the discipline as whole and about what matters to different generations, for instance, the increase of dissertations on climate change and on climate change activism. However, besides *what* is being studied by students, it is also fascinating to explore how these choices in topics and themes are justified and valued. Such frameworks are sometimes implicit and other times explicit: but, generally, over time they became more explicit, with chapters dedicated to conceptual or theoretical frameworks, or at least references to such frameworks in either a literature review or a methodological chapter. As with the likes of Kitchingham's dissertation mentioned above, particularly from the late 1990s there is an explosion of such frameworks being laid out in the dissertations, but the scope and treatment of this in dissertations were still diverse: where some students might merely name-drop a few predecessors and their concepts, others would offer sustained critical engagement with bodies of thought. By the mid-1990s, such frameworks become part of virtually every dissertation. By placing their own research in such frameworks, whether cast as discipline-wide or set within a particular subdiscipline, students explain why their research is 'worth it', both in an epistemological and in what might be termed a political perspective. The absence of these frameworks before the 1990s does not mean that students from earlier cohorts were entirely unaware of their academic allegiances, but for later cohorts they tend to become explicitly mentioned.

After his arrival in Glasgow in 1995, Philo put together the aforementioned core course 'Geographic Thought', focusing on the history of geography, particularly its older conceptual orientations, as well as various contemporary theoretical dimensions of geography. In an interview, he reflects on starting this course in the mid-1990s:

> There was a need [for] a stronger academic scholarly basis. I was brought in as a professor with the mission, to some extent, to change this. It wasn't me alone, but I did do quite a lot of the work of revising how we taught students. I was particularly supported by Jo Sharp and Paul Routledge, two critical human geographers. I put together the 'Thought' course, that still continues today. I took over a previous core course called 'The Evolution of Geographic Thought' – the phrase 'evolution' being telling,

suggesting a somewhat banal, gradualist approach. It had very little content, and Glasgow students, when I arrived, had next-to-no grasp of disciplinary history, conceptual development, or even how literature and concepts might be deployed to frame empirical inquiry. In fact, students struggled to name geographers or identify readings that influenced them.

(C. Philo, interview, 2019)

Again, changes in academic staff, leading to changes in the curriculum, very much transformed the knowledge and understanding of how students might relate their own work to that of predecessors and contemporaries, and also how they might write down these justifications and relations.

In histories of geography, regional geography is often pictured as a conceptual approach that became out of fashion after the 1940s. However, in Glasgow's dissertation collection, regional geography is still very much present until the mid-1970s. The mid-century decisive break was thus not as decisive in the local context of the Geography degree at the University of Glasgow. What is interesting to note here is that, whereas regional geography was alive and well for a long time, there are fewer traces of environmental determinism in the dissertation collection. A concern for human–environment relations has definitely persisted, but with the emphasis usually on the two-way nature of the relationship. In the following subsections, four of these conceptual spaces will be discussed in detail: regional geography, spatial science, Marxist geographies, and humanistic geographies, followed by a succinct overview of some other conceptual frameworks. They all have a very striking similarity: namely, there is a delay of approximately 20 years from their timeline in 'standard' narratives about history of geography (e.g. Philo, 2008 and Johnston and Sidaway, 2016) and the rough image of the timeline that emerges from the analysis of the undergraduate dissertations. This evokes questions about how much – or little! – the undergraduate curriculum is subject to change, and to what extent such change relies on individual staff members, either newly hired or doing innovative research. The analysis in this chapter thus pivots between the 'overall' timeline of conceptual approaches in geography and the local conceptual space of geography at the University of Glasgow.

Regional geography

In standard timelines or narratives about the history of Anglophone and European geography, regional geography is often connected to the first half of the twentieth century (e.g. Philo 2008: xii; Livingstone, 1992), supposedly supplanted by a massive shift towards 'spatial science' from approximately 1950 onwards. Pre-war, 'chorology', meaning the study of the region, was the 'cardinal principle', argued for by, for instance, Hartshorne (Livingstone, 1992: 309). Although the region still retained its centrality in geographical inquiry over time, different approaches in geography, especially spatial science, replaced the idiographic analysis of regions with nomothetic 'hard science'-like geography.

This did not mean that the region as a 'scale of inquiry' disappeared in later studies or that regional geography was totally gone, but that it was not the key concept of geographical research as it had been before. The reduced dominance of the study of the region in the dissertation is thus to be expected from the mid-twentieth century. The appearance of 'systematic dissertations' clearly reflected wider debates about the increasing emphasis on 'systematic geography' as opposed to ' regional geography': in a systematic geography dissertation the student-researcher concentrated systematically on the geographical dimensions of a specific phenomenon such as plants ('biogeography') or the economy ('economic geography').The decrease of regional geography and the increase of systematic geography hence closely related to the identification, or development, of different (systematic) subdisciplines.

The 'regional dissertation' was, until the mid-1970s, not an option but a requirement for geography students at Glasgow University, was a place where human and physical geography met in the educational context, perhaps representing the strongest 'unity' in the geography curriculum of all decades studied. Some later dissertations still explicitly operate at and even name the regional scale in their research, to be sure, yet the 'nature' of such dissertations is very different from that found in the classic regional dissertations, consistent with general reflections on the wider place of the regional in geography. The evident follow-up question is why, when regional geography slipped out of fashion in the 1950s, students in Glasgow were still required to do a significant amount of work as regional geographers right through to the 1970s. Ian Thompson, Professor as well as Head of Department from 1976 to 1986, might have been a big influence on this continuation of the regional dissertation, given his specialism in the geographies of France and debt to Vidal de la Blache's regionalism, but the 'conservative' attitude of the department was widespread. Reflections and memories of former staff members help in interpreting this continuing importance of regional geography in the 1970s at the University of Glasgow. In the 2009 special edition of the *Scottish Geographical Journal*, celebrating the centenary of Geography as an institutionalised presence at the University, Gordon Dickinson discusses regional geography:

> At the start of the 1970s regional geography was still regarded as a major strength of the Department in both teaching and research. In the discipline's search for stronger conceptual bases, the kind of synthesis that characterised much mid-twentieth century regional geography was held in high regard at Glasgow. This did not mean that quantification or the development of more philosophically based theory was ignored. Topographic Sciences helped develop the former within geography, whilst the latter saw innovation in the work of a number of the younger staff, especially in human geography.
>
> (Thompson et al., 2009: 334)

The disappearance of the regional dissertations in the mid-1970s is perhaps just a very late follow-up of the disciplinary trend moving away from regional geography to spatial science; but, as Dickinson expresses here, although spatial science was perhaps an attempt to connect to the natural science disciplines in academia, this move can also be seen as something that brought even more 'problems' to geography. If regional geography 'pulled it together', what is left when you end up with the opposing pillars of one discipline, with the evident bridge taken away? Disciplines grow apart, slide out of fashion, are split up into multiple specialisations that grow as independent disciplines, and such moves are of course mirrored in undergraduate degree studies as well. Although the regional dissertation disappeared in the mid-1970s, the regional tradition indeed did not disappear as a whole, neither in Anglophone Geography nor in Glasgow's undergraduate curriculum. Whereas the number of dissertations combining human and physical geography declined with dropping the requirement to do a regional dissertation, the regional tradition, perhaps slightly out of fashion, was still one of the ways to pursue disciplinary unity after the 1970s, although doubts about the 'artificiality' of this unity remained present.

Spatial Science

In disciplinary timelines, spatial science is sandwiched by two decisive breaks, being a fundamentally different framework from the ones preceding and the ones following later, and this of course is because the underlying presuppositions of what geographical knowledge is, and what geographical research should look like, fundamentally differed between spatial science, its precursors, and its subsequent antagonists. The first explicit spatial science dissertation in the archive arguably dates from 1965[1] and the second one not until 1971,[2] while this approach, in its typical vein as introduced in the 1950s and 1960s, is traceable until the early 1980s. Quantitative research played a significant role in the research of many student-geographers in the 1980s and 1990s, although the distinctive spatial science emphasis on 'locational analysis' is not noticeable in such quantitative research projects, indicating a crucial difference between spatial science dissertations and the general use of quantitative methods. One example of a spatial science inquiry is Rosemary Brooks' dissertation *A study of spatial patterns of movement in the villages of Balfron, Killearn and Blanefield in relation to their varying distances from Glasgow* (1974). Brooks quantified the movements of residents of three villages based on a questionnaire, and the aim of her research was formulated as follows:

> By examining the bases for, and patterns of, spatial interaction we move a step closer towards an answer to one of contemporary Geography's most pertinent questions. 'Why are spatial distributions structured the way they are?'
>
> (Brooks, 1974: n.p.)

The 'why' in her research question was perhaps overstated, because the dissertation itself barely discussed an answer to the 'why' and simply emphasised the 'linear continuum' of the three villages because of their relative distance from Glasgow. Although many later dissertations asked similar research questions, the approach here was slightly different because it did not include any other variables (potentially causal factors) that might be influencing residents' mobility. Many of the spatial science dissertations in the collection are part of settlement studies: clearly, therefore, some conceptual frameworks tend to be exercised within specific subdisciplines. Settlement geography is itself a somewhat split subfield: varying from descriptive studies researching environmental factors influencing settlement siting and growth to spatial science research aiming to measure and prove law-like postulates underpinning the likes of Christaller's Central Place Theory.

Overall, the number of obviously spatial science dissertations in the archive is small, and different from the lagging timeline of regional geography in Glasgow compared to the wider geographical tradition; indeed, there is no similar significant delay in the drop-off of spatial science dissertations after the heyday of such studies in the wider discipline, partly because it seems as if spatial science – certainly in any developed, self-aware manner – never gained a massive foothold in Glasgow. It is difficult to find an explanation for this situation, but it might be that, with the continuing centrality of regional geography in the curriculum, there was a conscious or even unconscious aversion towards spatial science and its demands in terms of quantification, statistical analysis, and maybe modelling. Alternatively, spatial science 'enthusiasts' – those students who might favour a more quantitative and 'technological' approach to geography – were maybe drawn to the Topographic Science degree that the department ran from 1964 to 2004. The emphasis upon regional geography was probably caused by staff preferences for such a form of geography, which is incompatible with the fundamentally different assumptions and expectations of spatial science that emphasise the 'law finding' character of the discipline rather than the particularising character of regional geography. Although hard to substantiate here, there is evidence to suggest that Glasgow staff of the 1960s–1970s were averse to, even afraid of, spatial science in anything but its more low-level and descriptive – spatial pattern-describing – manifestations (Philo, 1998). Not all of the central elements of spatial science disappeared entirely; some were later reinvented in the context of GIS and geospatial-computational methods, allied to the rise of 'big data', entraining both human and physical geography. GIS and automated cartography have remained and arguably become more prominent in many physical geography dissertations in Glasgow over time.

Marxist geographies

The first mentioning of Marxist theories in the sampled dissertations can be traced to a social geography dissertation from 1986, entitled *An analysis of inequality in pre-school provision* (Allan, 1986). This emphasis on inequality makes its appearance in social geography dissertations in the late 1980s. Margo Allan explicitly mentioned that geography had neglected the problems of preschool provision and the role of 'territorial justice and injustice' in these matters, referring to 'contemporary neo-Marxist theories' (Allan, 1986: 33) and social theories on resource allocation. Overall, this dissertation was highly political, politicised even, and Allan paid a lot of attention to the theoretical and conceptual framing of her research. It is interesting to note that this dissertation is also a very early example of a dissertation explicitly acknowledging young children as a social group worth studying, hence also comprising an early brush with an 'outsider' group. Besides the dissertations explicitly mentioning Marxist thought, there were some earlier examples of dissertations that arguably fitted into the Marxist tradition: for instance, Irene Hair's dissertation *The geography of riots in Glasgow and Lanarkshire, 1770 to 1850* (1978). This dissertation, clearly a historical geography project, addressed questions of riots, resistance, and radicalism:

> During this period the Scottish lower classes became increasingly aware of radical political ideas – as seen by the increasing incidence of industrial and political violence; However radicalism seemed to have had less impact on the Scottish population than on the English – probably because, unlike England, Scotland had no marked constitutional element in its history. The price of bread, religion and the availability of employment were more likely to cause riots in Scotland than demands for universal suffrage.
>
> (Hair, 1978: 3)

Research projects such as Hair's demonstrate that it is not always easy retrospectively to impose a certain conceptual frame on the work undertaken, especially when it is not explicitly mentioned. Hair's dissertation is certainly a historical–political piece of work, but this is more an indication of what *subdiscipline* it involved: neither historical nor political geography need necessarily be Marxist geography at all. However, the specific themes, emphasising radicalism, inequality and political activism, might indicate at least some 'light version' of Marxist ideas. Unfortunately, Hair does not mention her supervisor or elaborately discuss any literature. Glasgow's Geography Department did employ academic staff during the 1970s that taught radical and Marxist geography. Ronan Paddison and Allan Findlay published on the contexts of theoretical–historiographic geography (e.g. Paddison and Findlay, 1985) and Paddison also wrote about historical materialism in Glasgow's student-led journal *Drumlin* (Philo, 2020). As Hair's dissertation implies, there was at least some attention to such approaches, but it was not until the mid-1980s that dissertation-writing students convincingly presented Marxist geography dissertations.

The conceptual space of Marxist geography is recognised in the Glasgow dissertation archive significantly later than it is in human geography as a whole. In the late 1980s and during the 1990s there are still relatively few students referring to (neo-)Marxist literature or concepts, and it is much later, during the 2000s and 2010s, that the study of inequality becomes more prevalent with references to Marxist geographical researchers such as David Harvey. The socio-political dissertations of these years often provide in-depth theoretical and conceptual justification for numerous highly politically charged approaches. This development, as well as the relative 'lateness' of it is not only a very obvious reflection of, again, the roles of individual staff members but also indicates a bigger shift, further and further away from the spatial science conception of human geography with its emphasis on the scientific method, positivism, and objectivity. Furthermore, the sense became more palpable of geographers being able actually to *impact* the society that they are researching, and the increase of dissertations with such a leaning demonstrates that it must be one welcoming to more recent generations of geographers, again, acknowledging the uncertainty of what came first, a student population more interested in these matters or a curriculum that emphasised certain approaches.

Humanistic geographies

What might be judged as the characteristic concerns of humanistic geographies – peoples' senses of place and community; matters of perception and behaviour, where treated qualitatively not quantitatively – do feature regularly in older undergraduate dissertations, but not at all in any obvious, explicitly conceptualised way before the cohort of 1998. Humanistic geographies often go hand-in-hand with the attention to 'new' societal groups, seen in research projects studying geographies of children, of elderly, and of the LGBT+ community, a development clearly connected to changes in the curriculum. Humanistic and post-humanistic dissertations are also numerically 'high performers' in the scope of student discussions of such frameworks. Here, two dissertations from 2002 will be discussed as examples. In his dissertation, *The effects of in-migration on sense of place in the small rural town of Castle Douglas* (2002), Dark structured his literature review under four themes: the concept of place, population movement and rural change, community and home, and 'centredness'. On the first page of the dissertation, he presented some reflections on these key concepts:

> In this paper it is the underlying idea that places are constructed and are given meaning. Every place, whether it is a home or a town, is understood in common terms by those who use that space and interpreted in differing terms, by others who are observing that place as outsiders. This duality is described by P. Knox (1995) as the 'betweenness' of place.
>
> (Dark, 2002: 1)

The reassessment – and alertness to the heterogeneity – of the concepts of space and place evoke strong associations with postmodernist thought, not uncommon for human geography dissertations from the early 2000s. Dark refers in his review chapter to rural geographers such as Ilbery (1998), but also to more conceptual work of geographers such as Matless and Buttimer.

Another clear cut at humanistic geographies is found in a later dissertation written by Chris Dickie: *The wonder years: Imagining the spaces and places of adolescence* (2006). His research emphasised memory, imagination, and embodiment as key concepts:

> My general aim then is to provide an insight into the personal, and shared geographies of adolescence, not through means of an 'intrusive' ethnographic study of a group of adolescents, but rather through an exploration of the memories, reveries and imaginings of a few twenty something's [sic] who in a real sense were that 'other' in the not too distant past. My study in turn, will hopefully help hint at an understanding of the ways in which we use our bodies to fit in to the world, both in terms of culturally accepted categories of identity (childhood; adolescence; adulthood) and in terms of less obvious, homely, ordinary types of association.
>
> (Dickie, 2006: 2)

Dickie's dissertation also hinted at phenomenological elements, referencing 'bodies fitting in the world' and with aspects derived from non-representational theory, which was then less than a decade old as named approach within the discipline's academic literature. He referred to 'going beyond representation', 'exploring the full humanness', and research methods alive to the possibilities and insights of the human body as well as to the limitations. In the most recent cohorts that were part of my sample, there are many such dissertations, although Dickie lent relatively more attention to the building up of an appropriate conceptual framework. The ideas and approaches associated with post-humanistic geographies – retaining an interest in the likes of emotions, affects, embodiment, and the 'vitalities' of being alive, but displacing the figure of the autonomous, self-controlling singular human being – are perhaps even more recognisable in the dissertations, with an even more explicit usage of conceptually charged words and sentences than was true for the dissertations cleaving to earlier humanistic geographies.

Other conceptual frameworks

By only discussing regional geography, spatial science, Marxist geography, and humanistic geography many other conceptual frameworks have been ignored. I provide here brief examples for three further conceptual frameworks: feminist geographies, postcolonial geographies, and non-representational theory.

The first dissertations focusing on women as a distinctive social group are found in the cohort of 1992. Such studies take sex and gender, both basically taken as biological givens, as a 'marker of identity', similar to, for instance, ethnicity, nationality, or socio-economic 'class'. It is not until the cohort of 1997 that geographies of women really seem to take off as a conceptually inflected focus of inquiry, with a substantial number of dissertations in the late 1990s taking gender as a concept and relating it to a variety of foci of inquiry: for instance, rural economics and politics. Such studies addressed women, or gender, in relatively 'traditional' geographical studies, focusing on women as a social group, but are not yet building upon critical feminist ideas. This novel focus of inquiry in itself, though, already embodied criticism of how women were overlooked in the discipline before, hence showing disciplinary awareness. These examples of work on geographies of women have been followed by later projects emphasising two other themes: gender and identity or representation of identity; and gender and public space, the latter including many projects on gendered dimensions of danger, fear, and the perception of fear. Unlike the earlier examples, these dissertations sometimes engaged explicitly with the 'gender division of space', critically discussing gender inequalities in access to and utilisation of different kinds of physical (or even virtual) spaces. There is also one specific subdiscipline for dissertation work in which geographies of women have been markedly present, development geography: such dissertations were mostly written in connection with fieldwork expeditions to Tanzania organised by staff members John Briggs and Jo Sharp. The appearance of dissertations on geographies of women in 1992 seems relatively late in relation to what was happening on the discipline's research frontier, but, as already discussed above, this delay is not uncommon. The sudden increase towards the end of the 1990s might also be explained by the hiring of Sharp, a leading feminist geographer herself, appointed in the mid-1990s and having an increasing influence on the department's curriculum as the 1990s progressed into the 2000s. Indeed, many of the dissertations from the later 1990s and into the 2000s covering feminist studies and geographies of women were supervised by Sharp. The increase of feminist geographies from the late 1990s also went hand-in-hand with the rapid rise of female geography students in Glasgow. Although the Glasgow geography degree from the start was admitting a mixture of male and female students (in the 1960s the percentage of female students was around 35%), towards the end of the century women, especially ones taking human geography options, started to outnumber the men. And, indeed, all explicitly feminist geography dissertations encountered were written by female students!

Literature from the field of postcolonial geography is, unsurprisingly, referenced in some development geography dissertations, but in the dissertations *not* part of the Tanzania research trips mentions of postcolonialism as a conceptual framework have been scarce. There are two critical postcolonial dissertations obviously identifiable, one cultural geography dissertation on postcolonial representation and one historical dissertation on imperialism. The organised expeditions to Tanzania, as well as to Egypt, will likely have attracted students in those cohorts who were interested in this school of thought, but it is still surprising that in the many cohorts when such expeditions were absent the entire cohort seems to ignore postcolonial thought and, more specifically, ignores continents other than North America and Europe. The cohort of 2006 includes one impressive example of conservation geography that draws upon the value of indigenous knowledge, entitled *Land, culture and conservation: A study of indigenous development within the Maori communities of Whakaki and Waimarama* (Donald, 2006). In this, Michael Donald investigated one conservation project from a postcolonial perspective:

> The true empowering nature of the Nga Whenua Rahui projects at Waimarama and Whakaki is predicated upon combining conservation practices with traditional Maori systems of land rights and knowledge.
> (Donald, 2006: 35)

This project seems to be unique in the dissertation archive. It can be concluded that the expeditions organised for several cohorts were indeed a fundamental contribution to the broadening of horizons – worldly, conceptual, and ethical-political – of student-geographers.

As mentioned in the section on humanistic geographies above when discussing Dickie (2006), discussions of NRT are often followed by references to phenomenological literature and vice versa. A second example is found in 2014, in the dissertation *The perambulatory geographies of Iceland women in Reykjavik* (Halliday, 2014). In this, Kirstin Halliday adapted a phenomenological approach, interviewing participants while walking and asking participants to create a 'time-space emotion' diary. In her dissertation, she referenced the work of anthropologist Tim Ingold, who is often aligned with non-representational thinking:

> ... within non-representational theory and its phenomenological foundations, walking practices are fully accounted for by the accumulation of lived embodied knowledge through environmental immersion, rather than viewed as being resultant from a higher social world. Tim Ingold's theorisation of walking as a way of thinking reflects this assertion of the grounding of social life within lived walking practice (Ingold, 2004).
> (Halliday, 2014: 8)

The emphasis on embodiment and also practice, the embodied practices of undertaking particular activities or tasks, increased from the late 2000s, which might be attributed to the hiring of Hayden Lorimer a few years earlier. Other examples of the influences of NRT found in the dissertation archive surface in geographical studies of creative disciplines and forms of art.

Overall, these 'waves' of appearances and disappearances of conceptual frameworks appear strongly connected to particular members of staff as well as to fieldwork opportunities and curricular changes. This point indicates that, although students in their dissertations undertake research independently, staff and curricular cannot but influence what student-geographers recognise as potential research foci and methodologies, and thus what 'kind' of geographer they want to become. The wide variety of conceptual frameworks *within* some cohorts indicates that geographers-in-the-making discovered what they actually found fascinating, interesting, and worth studying, which can only have made their research experiences more positive. Learning about different conceptual frameworks and different bodies of literature is thus a fundamental aspect of becoming a geographer.

Conclusion

Narrating the history of geography as it has been taught in Glasgow seems to make a very strong distinction between, on the one hand, regional geography, and, on the other hand, 'systematic' geography. The continuation of the requirement to produce a regional dissertation, up to the 1970s, indicates that there was a strong regional tradition in Glasgow. There was a long transition period towards more demarcated and specialist 'systematic' knowledge production during that decade, and, together with this shift of emphasis, expectations from students to pay more attention to research design also started to grow, if slowly. It was not until the end of the twentieth century that the presence of a strong engagement with academic literature arose, with conceptual framings and a strong awareness of how a student-geographer as a geographer-in-the-making might tweak disciplinary or subdisciplinary traditions, maybe by studying something new or studying something in a new way. So, instead of demonstrating a certain set of skills that proved students were 'real geographers', the students often aimed to affect a small 'change' in the universe of geography. Sometimes these practices of identifying a gap in the literature were more or less an echo of what was required of them, like a form-filling exercise – refer to some academic studies, identify a gap, reveal 'something novel' – but a majority of students seem seriously to consider these choices. Such a consideration asks for a self-awareness that is rarely explicitly found in the earlier cohorts. From the late 1990s, students were aiming to identify that 'gap' in the literature and tried actively to *add* something novel to existing knowledge.

This narrative is a very broad one, then, rushing over the valuable contributions of apprentice geographers from different cohorts. Dissertation research is not necessarily inferior to that underpinning other academic publications, despite the status of its practitioners, but its primary purpose is very different. For the wider academic community, therefore, emphasising solely the procedural–educational aims of the undergraduate dissertation as a curriculum requirement does risk substantial amounts of (in some cases, really quite impressive) knowledge remaining behind closed doors. For instance, the discussed dissertation by Katy Kitchingham, *Investigating 'nature' – A case study of Crawford Lake Conservation Area in Ontario, Canada* (1998), distinctively connects together questions concerning conservation, tourism, gender, and ethics of land ownership in a colonial perspective. Although human geography was emphasised in this chapter, the narrative is quite similar for physical geography projects. The emphasis in the latter might be a little more explicitly on the application of techniques and skills, and such dissertations reveal an increasingly critical attitude to what specific methods can and cannot achieve, but here too there are signs of the student-physical geographers increasingly addressing the 'whys' and 'hows' of the work that they and their fellow physical geographers are doing. As noted at the outset of this chapter, whereas some students use 'the tradition' as justification, others emphasise their 'innovativeness' as the value of their work, with some suggestion of a tendency over the period from the former to the latter. These individual differences might be connected to students' academic self-assurance, their own inspirations and aspirations included, but also to the relationship with supervisors and curricula (and how they mutate).

This chapter demonstrates, once more, how entangled the narrative becomes. Dissertations tell us something about the department and its status and readiness to embrace differing disciplinary and conceptual possibilities. A changing student population in a changing department, part of a changing university in a changing society: it is almost surprising to find so many ideas and concepts running in a shared fashion like a thread through the cohorts. Disciplinary awareness as well as self-awareness are more comfortable parts of dissertation for later students, but, as seen in the opening citation of this chapter from Sheila Munro, more than 70 years ago, students – or at least some students – also thought about what they 'should do' – and, crucially for this chapter, 'should think' – as geographers.

Intermezzo 5: Medical/Health Geographies

Dissertations about health are discernible throughout the different decades in the archival collection, but there is a clear shift concerning the approach taken by such health-related studies by students. Generally, the main focus of medical geography studies by students before the mid-1990s split across three elements of health care: access, logistics and spatial variations. Such studies often connected health to social-urban questions related to poverty, class and the urban environment. In her dissertation entitled *Spatial distributions of deprivation and children's heights in Lanarkshire* (1994, see Figure 5.3), Frances Murphy interrogated the relationship between poverty and health, focusing not on health defined by the 'absence of disease' but rather by another 'representation of general health', height:

> The main purpose of this study is to discover whether or not there are spatial patterns in children's heights across Lanarkshire, and how these are linked to deprivation levels. Height was used as a representation of a child's general health, and the indicator of deprivation used was the number of children receiving free school meals.
> (Murphy, 1994: 4)

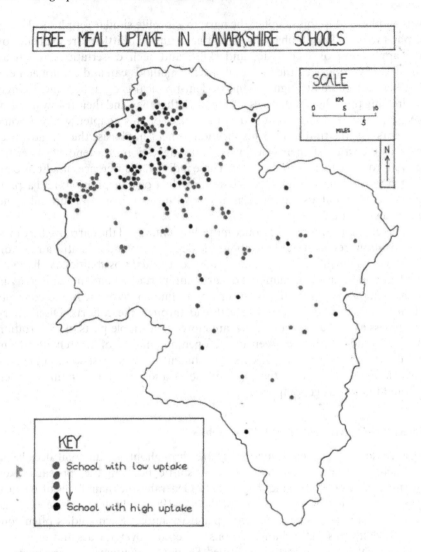

Figure 5.3 Map of free meal uptake in Lanarkshire schools (Murphy, 1994).

Her methods were mainly quantitative, combining school medical records from the Lanarkshire Health Board and statistics showing the uptake of free school meals to relate 'deprivation' to 'height'. She explained why studies like these are important:

> This topic is of interest at the present time because of mounting evidence of an increase in inequalities between the richest and poorest segments of the population, with a corresponding rise in health inequalities.
>
> (Murphy, 1994: 4)

She also reflected on the subdisciplinary field of which her research might be judged a part:

> The field of medical geography has been especially concerned with describing spatial distributions of disease and attempting to explain why these variations occur. .... However, very little research seems to have been directed towards areal variations in other aspects of health, less drastic than disease.
>
> (Murphy, 1994: 6)

Murphy's dissertation can be seen as a clear example of the research done in the field of medical geography. The *Dictionary of Human Geography* provides the following description of medical geography, which clearly acknowledges the role of spatial diffusion:

> Medical geography is concerned with a variety of health and illness topics and is very much multidisciplinary in nature. Nutrition, communicable diseases, spatial diffusion, economic development processes, chronic, or lifestyle diseases, disability, violence, substance abuse, environment – health relations, healthcare systems and philosophies, location of health service facilities, and allocation of those in need of care to sources of care – all have geographic components.
>
> (Earickson, 2009: 9)

From the mid-1990s on, the connection between health and social-urban geographies becomes less present in the archive, and the theme of health is more often connected to research concerning development geography or hazard studies. The connection between flooding and flood risk and (public) health is more than once researched by students, echoing claims made earlier about the influence of Rhian Thomas's teaching about hazards, flooding and health, while there are many examples of 'health and development' dissertations. Some of the latter focus on a specific disease, while others take an approach that connects health to socioeconomic matters (somewhat similar to Murphy's 1994 dissertation mentioned above).

In her dissertation *The impact of disease on the socio-economic structure of the household in Dar es Salaam, Tanzania* (2010), Sarah Scholes explored the effects of disease at several levels: 'both at a household level as well as from a global perspective' (Scholes, 2010: 3). She identified a 'research deficit':

> The fact that the impact of disease on the household tends to be taken for granted is evidence by the absence of significant research in relation to places such as Dar es Salaam. As Steven Russell (2004: 143) explains, more research is needed on the household costs of disease, how the household responds to disease, and the extent to which disease exacerbates poverty. This project seeks to readdress that research deficit.
>
> (Scholes, 2010: 3)

Scholes then stated the following research objectives:

> To establish the most common diseases and the extent to which they affect the socio-economic structure of households in Dar es Salaam; To establish how the members of the household 'live' or cope with disease; To analyse the effects of the loss of the economically active parent in the household and how this affects the family dynamics of the household.
> (Scholes, 2010: 10)

These objectives clarify how Scholes aimed to combine quantitative and qualitative research methods, striving to connect medical data, socio-economic data, and behavioural and emotional data with each other. This emphasis is not so distant from Murphy's 1994 dissertation on physical health, and it is one recognisable throughout the full dissertation archive. There are studies here that address *mental* health as well, reflecting a 'niche' strength of the Glasgow department, especially in the 2000s and 2010s. Many of such dissertations become reconfigured as contributions to the social geographies of 'outsiders'. There are, for instance, several dissertations on the histories of mental health institutions. Overall, medical geographies or health geographies are traceable throughout the archive but the subdiscipline has nonetheless remained a relatively small focus of study.

Notes

1 The pattern of central places north of The Caledonian Canal (Munro, 1965).
2 An investigation into the relevance of the 'gravity model' and related hypotheses for the purpose of explaining the volume of interaction between urban settlements in the Scottish counties of Fife, Kinross, and Clackmannanshire (Urquhart, 1971).

Bibliography

Allan, M.K., 1986. *An analysis of inequality in pre-school provision*. Undergraduate Dissertation, University of Glasgow.

Ambrose, J., 1970. *Twelve parishes in the Welsh Borderland; Regional planning problems in an area of transition*. Undergraduate Dissertation, University of Glasgowe.

Black, H.M., 1998. *Juvenile delinquency – A class based phenomenon?* Undergraduate Dissertation, University of Glasgow.

Brooks, R., 1974. *A study of spatial patterns of movement in the villages of Balfron, Killearn and Blanefield in relation to their varying distances from Glasgow*. Undergraduate Dissertation, University of Glasgow.

Cresswell, T., 2013. *Geographic Thought: A Critical Introduction*. Chichester: Wiley-Blackwell.

Dark, C., 2002. *Incomers to the food town: The effects of in-migration on sense of place in the small rural town of Castle Douglas*. Undergraduate Dissertation, University of Glasgow.

Dickie, C., 2006. *The wonder years: Imagining the spaces and places of adolescence*. Undergraduate Dissertation, University of Glasgow.

Domosh, M., 1991. 'Toward a feminist historiography of geography'. *Transactions of the Institute of British Geographers*, 16(1), 95–104.

Donald, M., 2006. *Land, culture, and conservation: A study of indigenous development within the Maori communities of Whakaki and Waimarama*. Undergraduate Dissertation, University of Glasgow.

Earickson, R., 2009. 'Medical Geography'. In: Kitchin, R., Thrift, N. (eds). *The International Encyclopedia of Human Geography*, 9–20.

Gaunt, R., 1998. *Glasgow 1999: A deconstruction of the educational programme surrounding a 'year of' event*. Undergraduate Dissertation, University of Glasgow.

Hair, I.N., 1978. *The geography of riots in Glasgow and Lanarkshire 1770 to 1850*. Undergraduate Dissertation, University of Glasgow.

Halliday, K., 2014. *The perambulatory geographies of Icelandic women in Reykjavik*. Undergraduate Dissertation, University of Glasgow.

Hastings, G., 1990. *Counterurbanisation at the periphery. Case study – Baillieston*. Undergraduate Dissertation, University of Glasgow.

Johnston, R.J., Sidaway, J.D., 2016. *Geography and Geographers: Anglo-American Human Geography Since 1945* (7th ed.). London: Arnold.

Kitchingham, K., 1998. *Investigating 'nature' - A case study of Crawford Lake Conservation Area in Ontario, Canada*. Undergraduate Dissertation, University of Glasgow.

Livingstone, D.N., 1992. *The Geographical Tradition: Episodes in the History of a Contested Enterprise*. Oxford: Blackwell.

Mauritzen, A.S., 1994. *The formation and morphology of scree slopes in relation to glacial downwasting of Sandfellsjokull in southern Iceland*. Undergraduate Dissertation, University of Glasgow.

Moore, C., 2006. *"They always shout at us": An investigation of adults' control over teenagers' recreational geographies*. Undergraduate Dissertation, University of Glasgow.

Munro, A.L., 1965. *The pattern of central places north of The Caledonian Canal*. Undergraduate Dissertation, University of Glasgow.

Munro, S.M., 1954. *Aviation as a new factor in geography*. Undergraduate Dissertation, University of Glasgow.

Murphy, F., 1994. *Spatial distributions of deprivation and children's heights in Lanarkshire*. Undergraduate Dissertation, University of Glasgow.

Paddison, R., Findlay, A., 1985. 'Radicals vs positivists and the diversification of paradigms in geography'. *Espace Géographique*, 14(1), 6–8.

Philo, C., 1998. 'Reading Drumlin: academic geography and a student geographical magazine'. *Progress in Human Geography*, 22(3), 344–367.

Philo, C. (ed), 2008. *Theory and Methods: Critical Essays in Human Geography*. London: Routledge.

Philo, C., 2020. 'Radical geography from the *Drumlin*: An academic commemoration of Ronan Paddison (1945–2019)'. *Space & Polity*, 24(2), 132–155.

Scholes, S., 2010. *The impact of disease on the socio-economic structure of the household in Dar es Salaam, Tanzania*. Undergraduate Dissertation, University of Glasgow.

Shiels, S., 1998. *The structure of deformation tills, within a contemporary glacial environment; Breidamerkurjokull, SE Iceland*. Undergraduate Dissertation, University of Glasgow.

Thompson, I.B., Dickinson, G., Lowder, S., Paddison, R., 2009. 'Recollections and reflections'. *Scottish Geographical Journal*, 125(3-4), 329–343.

Todd, T., 1990. *An investigation of the liverwort Herbertus Borealis*. Undergraduate Dissertation, University of Glasgow.

Urquhart, G.B., 1971. *An investigation into the relevance of the 'gravity model' and related hypotheses for the purpose of explaining the volume of interaction between urban settlements in the Scottish counties of Fife, Kinross and Clackmannanshire.* Undergraduate Dissertation, University of Glasgow.

Watson, G., 1994. *The effect of parent material on soil development; With reference to the geology around the Stonehaven area.* Undergraduate Dissertation, University of Glasgow.

6 Exploring the skills of geographers-in-the-making

Similar to decisions on fieldwork locations, foci of inquiry, and conceptual frameworks of dissertation research, the choices made by undergraduate geography students with regard to research methods were often inextricably connected to limitations in time and money, as well as to their personal preferences and skills. Such choices were also connected to the formal expectations and requirements of what a dissertation should 'be', 'look like', or 'do'. From counting pedestrians and cars, to formal and informal interviews and focus groups, and from taking soil samples to vegetation surveys: the methodologies deployed by Glasgow's undergraduate geography dissertations are diverse, and often within one dissertation several methods are combined. The *collection* of data is important and time-consuming, but the *analysis* of the collected data is just as important for the research to understand findings and to provide a valuable conclusion. Methods of data analysis have changed significantly over time, not only because of innovative technological and IT opportunities but also because of shifting disciplinary trends. The practical approaches found in the dissertation archive also have value in discussing epistemological questions. Furthermore, the methods used provide a very tangible sense of the grounded experience that student-geographers have while doing their research: how do they interact with their fieldwork location? How do they see their own role as a student-researcher?

Methods of data collection

The process of undertaking research projects is often presented as a linear process, covering consecutive phases such as establishing a research question or suite of questions, data collection, data analysis, and writing up the research. In reality, as any researcher will know, the process is usually a lot messier. Many student-geographers have combined activities such as observing, counting, measuring, and asking questions for their undergraduate dissertation.

Observing

The act of observation as research method sounds like a passive way of collecting data: the observing researcher, as an outsider, studying their animate or inanimate research subjects. Glasgow's geography dissertation archive demonstrates many of these 'observing outsider' forms of data collection, from observing material landscapes to observing human practices or behaviour. Whereas physical geography projects of later decades emphasise measurements and modelling, the art (or craft) of 'reading the landscape' was very prominent in these two decades. In 1966, Robert McAllister wrote his dissertation *Structure in the woodlands around Balmaha, Loch Lomond*. His research aim was to describe the forest structure in the woodlands around Balmaha. The observation and drawing skills of McAllister are very impressive (see, for instance, Figure 6.1).

Although students were aware of the 'subjectivity' of observation, it was still seen as an important method of data collection. Observation would often be followed by other methods, such as collecting vegetation samples, to *confirm* the observations, as Dickinson describes in his dissertation *Mountain vegetation in south-western Rhum* (1966):

> The results obtained from these samples were used, primarily to confirm the subjective observations carried out.
>
> (Dickinson, 1966: 2)

The observations and the sketches following the observations usually remained a substantial first step of collecting data and would also play a significant role in the writing-up and displaying of findings. In more recent decades, observation is still regularly (if less often than in the 1960s and 1970s) mentioned in combination with other methods of data collection, but in many of these cases the observation is phrased as a 'secondary' method. Although the dissertations thus seem to reveal a decrease of the centrality of observation as a method in physical geography dissertations, it seems that the fundamental shift is in the view of observation as itself a self-contained method, such that in later eras observation became merely a self-evident aspect of the fieldwork experience, not something needing to be discussed separately or elaborately. This change has consequences for the attention paid to and value of representing observations visually, by means of sketches and photos.

Observation is not exclusively a method of physical geographers since it is also used by many human geographers. Analysing the observations made in human geography dissertations display a big shift from observations of 'social landscapes' similar to the visual readings of physical landscapes, to a very different form of data collection by means of 'participant observation', not taking a 'observing outsider' perspective. This is research with ethnography – immersed inquiry in the social landscapes under research – as its main method, mainly used in social and cultural geography projects. For instance, in *The Faslane Peace Camp: Mobility and permanence in sites of resistance* (Wilbur, 2002), the student-researcher considered participant observation as 'the foundational basis' for his approach to data collection:

Exploring the skills of geographers-in-the-making   115

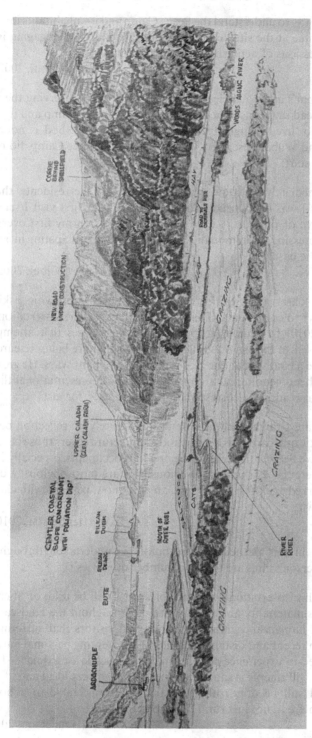

Figure 6.1 Landscape sketch: Loch Riddon (McAllister, 1966).

> ...the best way to understand the Peace Camp and its inhabitants involved spending time at the site, living as the residents do and engaging in the same expressions of activism as local participants.
>
> (Wilbur, 2002: 14)

As such, Wilbur's approach clearly entailed more than observing the inhabitants, and instead saw himself actively engaging with the Camp and its inhabitants, trying to 'live through' their experiences. This method is not without difficulties, and Wilbur described how people at the Peace Camp did not trust him from the start:

> I was immediately welcomed and introduced to the residents, though because as I was introduced as a researcher on my first visit I sensed a small degree of mistrust. I tried to eradicate this on my first overnight stay by expressing a commonality of purpose and participating in a small action at the base.
>
> (Wilbur, 2002: 16)

This form of 'observation' can be seen as a mixture of observing, asking, and 'living' as ways to collect data. This is also the case for the dissertation *Shenavall Bothy* (2010) by Stuart Henderson. He emphasises that attempting to immerse himself in the surroundings and events that are being researched led to a subjective narrative, but one that is revealing in many ways. He grounds his research in phenomenological arguments and non-representational theory literature, and also adopts a corresponding writing style, for instance:

> Buildings are straight forward: aren't they? Surely no reflection beyond the aesthetic or technical is needed and architecture covers those bases? ... We may give fleeting thought to the previous occupants, especially in our homes where the markings of previous lives are most obvious: a child's height marked against a doorframe, an adult now, perhaps with children of their own.
>
> (Henderson, 2010: 22)

Observation, in this particular dissertation, is undertaken through *all* the senses. Henderson's aims are to tell 'a number of stories':

> To write this dissertation a number of stories will be told: of Shenavall bothy – its materiality and reverberations throughout the landscape; of individual human and animal residents and visitors that surround the place both recent and distant; of a literal journey, my personal narrative of the 'lived-in' experience of being in this landscape for a time. ... Hopefully what will emerge is something that although very local and idiosyncratic will still be universally relevant and transferable to a host of different places – not just rural but urban or suburban.
>
> (Henderson, 2010: 3)

The examples discussed here demonstrate that observational methods for data collection are still very versatile: Wilbur and Henderson's observations go hand-in-hand with immersion, combining an outward as well as inward perspective. Such reflexive research means that the researcher is constantly aware of herself or himself and is questioning not only the specific focus of inquiry, but also how this affects oneself, and *vice versa*. Such reflexive practices are first mentioned in dissertations from the cohort of 1998 and become more common from 2006 onwards, itself a particularly clear indication of a changing discipline.

Measuring

In the 1970 dissertation *A geographical account of the Parishes of Coldingham, Eyemouth, Ayton and Chirnside, in the County of Berwick*, J. Graham describes the role of the regional geographer:

> Towards this end [helping the planner in determining the best future of the land in relation to the aims and values of the society concerned] the regional geographer may play his [*sic*] part in that, within his defined objectives, he assumes the role of field scientist who is concerned in principle to identify, measure, classify and describe the physical environment and the related socio-economic structures.
>
> (Graham, 1970: 3)

Seeing the role of the regional geographer primarily at the service of the planner is already an interesting perspective on the discipline, and not one that would likely have bothered Graham's counterparts doing regional studies from the 1950s into the 1960s. However, it is also interesting that Graham mentions a few verbs here: the physical environment and socioeconomic structures should be *identified, measured, classified*, and *described*. Some of these verbs mentioned are connected to data analysis (classifying, for instance), and others to data collection, such as measuring; and, indeed, measuring is central in many physical geography projects. Measuring is, like observing, a method found deployed by all cohorts.

One example of a dissertation using measuring as its main method of data collection is *An investigation into the relationship between beach morphology and beach sediment characteristics, with reference to the lochshore site of Cashel, Loch Lomond, Scotland* (Booth, 1994). Rachel Booth combined taking sediment samples by using quadrats with measuring different variables, such as beach gradient, wave height, wind direction, and wind speed. Booth pays a lot of attention in her write-up to describing the process of data collection and supports this by including several photographs of this process. She added a photo of her 'research assistant' (her boyfriend) demonstrating the use of the two-metre ranging pole used to measure wave heights. This, again, reveals the importance of students' social networks in doing their research: for some projects, it was even a necessity to get support from someone else, in carrying material or for safety measures. Booth thanks her boyfriend in the acknowledgements, giving an embodied sense of the methods being used:

> Mr Andrew Cowie must also be given special thanks for braving the cold water of Loch Lomond whilst collecting my offshore sediments and wave height readings.
>
> (Booth, 1994: 45)

A second example of the act of measuring as a means to collect geographical data can be found in *The effect of tidal variations on sediment concentrations and fluxes within an intertidal mudflat creek, Chichester Harbour, W. Sussex* (Keith, 1998). In her dissertation, Gillian Keith researched tidal variations within a salt marsh channel, with data collection taking place over five separate days within "a full fortnightly tidal cycle" (Keith, 1998: 9). Based on (mainly) water samples, Keith took multiple measurements: tidal current velocities, accretionary measurements, weather measurements, and suspended sediment measurements, combined with tidal data. Her dissertation is highly revealing in the practicalities[1] of taking accurate measurements:

> Initially it was thought that it would be possible to stand on the bridge to take measurements and use crates to stand on when the water rose above the bridge this however proved difficult and impractical. Therefore it seemed best to use a small rubber tender anchored on the bridge to take measurements from. All equipment could be stored on the boat and measurements could be recorded easily from this base.
>
> (Keith, 1998: 8)

To get these measurements right, there was lots of preparation work and consideration needed to decide on the best instruments and materials. Keith based the choices in data collection on methods found in academic literature, but also reflected on the support and help of her supervisor, as well as help with the equipment by supporting staff. Although 'measuring' might sound like a straightforward practice of collecting data, these examples demonstrate that to collect useful and correct data, there is a lot of intellectual as well as practical labour involved. A department possessing staff with enough expertise, time, and means made it possible for undergraduate students to experience, though small-scale, all stages of research and working with their own collected data, instead of working with, for instance, secondary sources (such as published tidal data).

Counting

The four ways of data collections discussed here are not mutually exclusive and often complement one another: for instance, many dissertations combine measuring activities with counting activities. However, there are also projects that solely focus on counting. One example is the dissertation titled *A geographical study of the Glasgow underground railway* (Parker, 1962). In this, M. Parker explores the 'traffic problem' of Glasgow, attempting to discover the achievements and limitations of 'the railway' – meaning Glasgow's subway. Parker does address the difficulties with their research methods in detail:

> At first, attempts were made to travel in the trains and count the number of passengers present in them in-between stations, and to estimate the numbers getting on and off at each stop. Alone, this proves an impossible task, even if only one coach was counted (the front coach assumed to carry an equal number of passengers as the back one).
>
> (Parker, 1962: 9)

Parker then continues with another method of data collection, described as an 'observation-procedure', but echoes a straightforward counting exercise, namely collecting subway tickets and counting these. This method of data collection was chosen by the student because it made comparisons of the data from different subway stations feasible. The data was then assembled and represented in multiple graphs, displaying the collected quantitative data in a legible figure, including the data on starting points and exits of subway journeys at a specific time. This counted data was subsequently analysed, leading to the conclusion "that there are two distinct parts to morning rush-hour, what might be termed the 'Worker's Rush' and the 'Clerk's Rush'" (Parker, 1962: 12), based upon the respective busyness of the subway to either city centre stations (the 'clerks') or stations near the shipyards (the workers) at specific times. Parker did not include any observations of the people he counted, so no remarks on, for instance, the gender composition of the subway users. This dissertation is a classic example of 'low-level' spatial science: quantifying simple space–time patterns of movement, without any sustained attention to the social content of what is being counted.

Such very clear examples of counting as the main method of research are found among many different cohorts but are most often found in economic and transport geography dissertations from the 1960s, 1970s, and early 1980s, counting number of shopping visitors or cars and other road users, and to a smaller degree in biogeographical projects, maybe counting birds. There is, however, also another 'form' of counting being done by a number of dissertation students, distinctive from the ones just noted which might be termed 'observational counting', and here is meant the counting of responses to closed questions as part of a large-scale questionnaire. Such questionnaires are not uniquely connected to a certain era, and are, without a doubt, the most used methodological 'instrument' for the Glasgow student-geographers over the whole time period of my study, either as primary research method or as a secondary, supportive, method. Collecting data from many respondents at one time might be seen as a relatively 'easy', low-intensive method now, but for the earlier cohorts it was often connected to long days of ringing doorbell after doorbell. Colin Thompson experienced this intensity, in his fieldwork for his dissertation *Analysis of the shopping facilities of Grangemouth* (1966). Thompson categorised the shopping facilities by type of shop, being a national, regional, or local 'type of establishment', and collected data on number of shops and floor space of shops. After this categorisation, he designed a questionnaire to survey Grangemouth's inhabitants' shopping habits. His methods were ringing 'random' doorbells, followed by the ensuing approach when the person refused. In the end, he collected answers from 1 in 30 households of Grangemouth, giving a total of 212, which seems like quite an impressive number. Thomson himself was not impressed though:

> The results of such a questionnaire are ... extremely interesting, and are valuable for explaining some of the shopping habits of the Burgh's inhabitants. However, the feasibility of stating that such a survey is representative of the town as a whole is questionable. In this case, only 1 in 30 of the town's inhabitants were interviewed. However, it is believed that the general unanimity of answers to most questions may make this survey, if not actually representative, at least reflective to a certain extent, of some of the shopping habits of the town as a whole.
>
> (Thomson, 1966: 16)

Such reflections are exemplary of prevailing ideas about what a questionnaire survey should 'do': ideally, the sampled cohort should be demonstrably representative of the target audience. Proving statistical representativeness is hard, especially for students using 'availability' surveys, rather than statistically controlled surveys. Many students, of all decades, formulate such limitations of their survey methods in their dissertations.

Dissertations from different decades provide ample examples of questionnaire surveys. Many of the questions in questionnaires from the 1970s and 1980s are closed, and aim at basic demographic profile data of participants (e.g. age and gender): such questions can be answered by ticking a specific box. Other questions require an answer on a Likert scale, for instance on one's level of concern about flooding. There are also many questions that require a numerical answer, about years of residence or years of working experience, or about number of visits to specific cities and sites. All of these forms of questions provide data that are easily counted and visualised graphically in some way. All these examples hence form a bridge between the data collection 'verbs' of 'asking' and 'counting': having data easy to count and quantify is found in dissertations from all cohorts, although, of course, many questionnaires also include open questions which can be used to collect examples and nuances, colouring, and complementing the quantitative data. Such open questions, soliciting more 'open-ended' qualitative responses, were rarer for the earlier cohorts using questionnaires in a 'social survey' guise – where lots of respondents and quantitative data are the goal – but became more common from the mid- to late 1990s, when humanistic geographies addressed the value of research of perceptions, ideas, and feelings, other than people as a source for 'factual' data. With such shifts in the focus of questionnaires, concerns about quantification came to matter less.

Asking

Questionnaires are of course pre-eminent instruments to *ask* people things, whether it is about their lives and living conditions or about their ideas and perceptions. Asking questions has always been central to students while doing fieldwork, and not all of these questions are asked in a formal situation but rather when talking to strangers when students are on unknown 'territory', seeking help with historical, contextual, and local knowledge of the specific place. Many of the regional geography projects seemed to rely heavily on these types of not-scheduled meetings with a variety of people from local communities, although, as already noted, methodologies are rarely explicitly discussed in any of these projects. Acknowledgement pages nonetheless reveal this reliance on existing community knowledge and also the anticipation of students that this kind of 'asking' is expected of them from their supervisors. Before turning to the 'classic' method of asking questions, interviews, it is revealing to look at the *other*-than-quantifiable questions briefly discussed in the section about counting. Towards the end of the twentieth century, more and more students not only included more open questions to questionnaires but also included 'mental mapping' exercises in their surveys: asking participants to draw maps of their own surroundings, whether these were maps drawn by children of their own neighbourhood or of nightlife consumers in Newcastle (Milligan, 2002). The student-researcher asked participants in focus groups to draw maps and use these materials to research the spatial awareness and spatial behaviour of the participants. She analysed the focus group by means of 'simple word count on single words' (Milligan, 2002: 13) to identify key themes and key issues.

Focus groups and interviews are to some extent similar but do hold important differences. Because of the number of 'interviewees' present in a focus group, the student-researcher is able to witness the 'intersubjective' discussions taking place among the participants: views, meanings, and experiences are shared, discussed, and perhaps contested. This might reveal more than the classic 'individual' interview, but also brings forward specific ethical as well as organisational challenges. Focus groups are regularly used by students from the cohort of 2002 onwards, and generally seem to be aware of such challenges. Orla Flanagan, for instance, used focus groups for her dissertation on the travelling community in the Northwest of Ireland (2006):

> the researcher was aware that although using a focus group facilitated a more comfortable and easier discussion participants maybe reluctant to discuss personal opinions in front of their peers.
>
> (Flanagan, 2006: 24)

In her dissertation, Flanagan explains the focus group format, consisting of a general introduction, the assurance of confidentiality and anonymity, explanation of the purpose of the session, an ice breaker activity, a group discussion of the experience of health services and suggestions for health services improvement. She decided on follow-up interviews with specific participants of the focus groups to discuss matters further.

Asking can be an activity that is non-scheduled, fluid, and flexible, able to move in different directions, or an activity that is very much scheduled and nailed down. Questionnaires are often discussed with supervisors beforehand, and the rigid form of questionnaires makes it easy to 'stick to the plan'. Focus groups ask for an active presence of the student-researcher, because group discussions can easily deviate to other topics and themes. It is up to the individual researcher to decide how much 'freedom' a participant, or a group of participants, has in deciding the direction of the conversation. Interviews, then, exist in many different forms: some are almost similar to questionnaires, whereas others are open conversations. Most interviews tend to be face-to-face, although there are many examples of interviews by mail, by phone, and online (both by using interfaces such as Skype and by asking questions through e-mail or chat). The face-to-face interview, though, remained the 'standard' form of interviewing throughout all cohorts.

Mixed methods approach

Many students from different cohorts have taken versions of a 'mixed methods' or a multimethod approach. Since the turn of the century students often expressly use this notion of deliberately 'mixing methods', especially in human geography projects, but combining methods was not uncommon for earlier students as well. Carol Welch, who went to Norway in 1974 to undertake fieldwork for her regional dissertation, wrote:

> Information was accumulated using written sources of information, personal contact with local people, correspondence with relevant bodies and through personal observation.
>
> (Welch, 1974: n.p.)

Archival and other documentary research, talking to people, and observing the physical as well as the social landscape often complement one other. Physical geography projects less explicitly mention this 'mixed methods' approach, but still demonstrate combinations of measuring and modelling, or mapping and observing, or any other combination of two or more ways of collecting data. Many geographers-in-the-making exhibit a variety of skills in the process of data collection, but bringing together these often very different data sets, requires skills in data *analysis*. Collecting data in the field often seems like a more adventurous activity than that of 'going home' and doing something with what has been collected, but it is in this transformation that the raw data might actually become knowledge, either 'new knowledge' or a (re)confirmation of existing knowledge.

Methods of data analysis

After collecting data in the 'field', whether this was from a landscape, a city, an archive, or any other form of fieldwork location, students returned to Glasgow, to the department or their own homes[2], and continued their project with the analysis of the collected data. The details of this process are of course different for every student and every project, but generally consisted of some or all of the following steps: structuring, categorising data, calculating, modelling data, connecting the data to the literature, and visualising the data in conjunction with writing up the dissertation as a whole.

Structuring, categorising, and calculating data

The practice of structuring and categorising the collected data is dependent on the kind of data collected and on the research objectives. Research projects that are quantitative in nature ask for different forms of analysis than qualitative projects. However, in almost all projects – some of the regional geography dissertations that are mainly descriptive skip this step, but in most other projects, including regional geography dissertations that go beyond description and use their regional study to offer some conclusions about 'regional unity' – there is a wealth of collected data that should be structured and categorised in a way. Some dissertations even have such a methodological categorisation exercise as their main focus; for instance, Ian Grieve's dissertation *A capability classification scheme for marginal land* (1970). In this, he uses the existing 'Land Use Capability Classification' (Bibby & Mackney, 1969), used by the Soil Survey of England and Wales, to test in the field whether it works to 'gain an estimation of the potential value of the land that is being classified' (Grieve, 1970: 13). Such a project is an exception in the dissertation archive: other students had to come up with their own structuring methods – their own schemes for ordering, classifying, and naming – and also, of course, their own justifications for choices made in the process.

Statistical analysis

Many students using statistical analysis as their main form of analysis mostly 'let the data speak for themselves' in their arguments, relying, usually with little comment or justification, on basic forms of tabulating, graphing, and simple descriptive statistics (measures of central tendency and dispersion) in order to present, summarise, and sometimes compare datasets. By the 1970s, though, there were students who explained that their own assumptions and ideas were influencing the decisions being made in determining and relating their variables. Jeff Collison wrote the dissertation *A geographical study of organised religion in an urban area* (1974), and in this, he explained that his selection of variables was guided 'by intuitive reasoning'. He arrived at a total of seven variables, for instance, the actual membership per congregation and the percentage of people over 65 in the neighbourhood of the church. These variables formed the basis for statistical analysis, which was executed digitally: "the computational work was to be performed by a computer" (Collison, 1974: n.p.). Though a funny quotation on first sight, this is one of the first dissertations that mention 'the computer' as a research instrument. Collison was supervised by Ronan Paddison, and it seems that many of the 'early adapters' of the computer were supervised by him. He, along with supporting staff members, is thanked in a number of acknowledgements pages for his help with 'computing' or 'the computer programmes'. In these early years, the use of computers was mostly limited to running standard statistical analysis programmes, whereas in more recent years they have obviously become the standard device for writing up the dissertation, as an access to data because of the internet and as a 'portal' to fieldwork locations when researching the virtual world, but also thanks to software packages for word processing, graphics, and, of course, quantitative analysis.

Another example of a student using computers for their data analysis was Alan Dowie, for his 1978 dissertation entitled *Cognitive mapping and the residential preferences of Drumchapel residents*. Early in his dissertation, he states his hypothesis:

> ... it is hypothesized that adult preferences will vary considerably from youth preferences, due to the differences in their level of psychological development and experience of the three spatial environments used as stimuli for expressing these preferences.
>
> (Dowie, 1978: n.p.)

Based on the answers given by residents when asked where in Drumchapel they would prefer to live, and by ranking their three most preferred and least preferred areas in Glasgow and Britain. In his dissertation, Dowie subsequently only discusses Drumchapel in the context of its desirability to other areas in Glasgow, so it is not evident why he also asked people about their most and least preferred areas in Britain as a whole. Dowie does, however, distinguish three *potential* scales to research the desirability of Drumchapel, and what the potential 'input' for any preferences on such a scale would be (e.g. direct experiences or images):

> At the local scale, Drumchapel can be perceived directly, although this may not always be the case with younger age groups, e.g. 12/13 year olds. At the regional scale, the city perhaps represents the largest area which can be perceived by way of direct experience by adults. ... At the national scale, where images, attitudes and mental maps, rather than direct perception or experience, play a more important role in forming any evaluation of distance places.
>
> (Dowie, 1978: n.p.)

Dowie developed a 'Desirability Index' for each area in Drumchapel. By using a cross-tabulation programme, he revealed popularity levels of these areas, the source for the visualisation displayed. Dowie's analysis was carried out by 'a post-survey coding of questionnaire responses for a computer crosstabulation programme' (1978: n.p.): seemingly a straightforward method to organise and analyse the quantitative data, whereas the data was *visualised* manually.

Especially from the late 1980s the use of more advanced statistical analysis became more common for undergraduate students. They also provide more details about the formulae used, often in appendices, and many students during the 1990s and early 2000s reflect on the use of Minitab. Mann–Whitney, Chi-square, and Correlation or Regression tests are the three most frequently mentioned statistical tests in the many dissertations that used statistical analysis between 1986 and 1994. The use of statistical techniques becomes so standard in human geography, a consequence of – and some might say 'hangover' from – the laggard diffusion of spatial science into university geography teaching from the later 1950s, that its 'standardness' is used as a justification by students:

> In recent years the application of statistical techniques has become commonplace in helping to solve problems of a geographical nature. As this is a geographical problem being studied in this paper, statistics shall be used in accordance with this trend.
>
> (Hastings, 1990: 27)

The 'obsession' with statistical analysis in these cohorts is a short but intense one. Already in the cohort of 1994 the engagement with statistical analysis has changed: it still plays a significant role, but not in almost *every* dissertation, and, if used, is more strongly embedded in existing literature. With the 'professionalisation' of physical geography in the department mostly caused by staff changes around 1994, the data analysis in physical geography dissertations projects changed as well. The statistical tests used became more advanced than the four mentioned above, often being combined with other forms of in-depth analysis, for instance particle shape analysis in the lab or deployment of modelling software such as Rocksoft.

Coding

In the same year there is also an interesting development perceptible in human geography dissertations. From 1994, the distinction between primary and secondary sources, already existent for earlier students, becomes explicitly addressed. This distinction saw students connecting their own research more and more to other existing academic literature, couched as 'secondary' sources, and using this literature as a starting point or as inspiration or example, for their own projects and 'primary' data collection and processing. It does not sound like an important shift, but by distinguishing primary data from existing source material, students began more actively to strive for 'triangulation', between data and literature, at the same time as becoming more aware of the value of connecting different forms of data to one other. The most significant shift in data analysis in human geography dissertations arguably occurred in the tangled realm of 'coding' practices, where 'coding' is a broad-brush term covering different ways of organising – and deriving themes and meanings – from qualitative research materials. Especially from 2002 onwards, students have displayed a very structured way of analysing large amounts of qualitative data, as collected in questionnaires, interviews, and focus groups or by participant observation and (auto)ethnographic research. Codes are decided on after 'immersion' in the data and are followed by counting and 'mapping' (or inter-relating) codes, as well as feeding into structuring relevant quotes and observations contained in a dissertation write-up. Coding, thus, is not itself solely a qualitative way of analysing data, but also one which can lend itself to some basic semi-quantitative analysis: for instance, as a basis for identifying, if not necessarily giving a precise value to, the prevalence of issues, themes or meanings in an overall body of qualitative research materials.

One example is found in Gavin Fleming's dissertation *Fear of crime: a case study in Glasgow* (2010). Combining the analysis of crime statistics with a questionnaire survey, Fleming explores several stereotypes that exist about crime levels in these areas. He describes his questionnaire survey as 'short interviews, in which detailed answers were provided', as opposite to 'quick tick' surveys oversimplifying the analysis of fear of crime. Fleming structures his coding per question: this allows his short interviews/questionnaire-style of collected data and objectives to add in some quantitative analysis of open-ended questions. Figure 6.2 displays the codes, code count, and related quotes for one of the survey questions.

The structuring of the codes demonstrates that, although there are a variety of answers given, the media is mentioned most often. Another example of coding is displayed in the figure below. Figure 6.3 demonstrates the *process* of coding: immersing oneself in the material, making notes, circling specific words and using colours and letters forming a provisional coding scheme.

Figures and appendices such as these two are included in all dissertations from 2006 that use interviews as one of their research methods, probably in reflection of explicit staff advice about how best students should show something of 'their working' in their written-up dissertations. The use of coding software such as NVivo or Atlas.ti by undergraduate students seems to be negligible: this might have been different if my study included some more recent cohorts than the 'Class of 2014'.

Besides coding there are other instruments and methods of qualitative data analysis such as narrative analysis or discourse analysis, but these are in the minority. It is important to note that these essentially depend upon some form of 'coding' exercise: key elements (being, for instance, issues, opinions, feelings, ideas) are identified and subsequently categorised and discussed. Coding is thus not a data analysis *method* in itself, but similar to the data collection verbs

Figure 4.1 - Hillhead Questionnaires, Coding of Question: *'Can you discuss why you feel you would be more likely to become a victim of crime in Maryhill than you would in Hillhead?'*

Themes/Codes	Count	Related quotes
Media	21	"you always see Maryhill in the media for the wrong reasons" "Newspapers are constantly reporting stabbings in Maryhill"
Outsider	9	"I wouldn't feel safe because I'm not from there" "most people feel comfortable when they're in a familiar place"
Youths	6	"the youths are a big problem in Maryhill" "there are so many gangs of youths there sometimes violent" "it is intimidating to see large numbers of youths roaming the streets drinking and causing bother"
Gender	7	"as a woman I do find that I sometimes feel intimidated" "I suppose I am always worried about rape and crimes like that yeah... I would think all women are"
Personal experience	3	"I have felt intimidated when visiting Maryhill" "I have a friend who was attacked there"
Physical environment	4	"you know it's more dangerous just from looking around" " I think the biggest difference between the Maryhill and here (Hillhead) is the look, like graffiti and there's much more rubbish I think"

Figure 6.2 Codes and examples (Fleming, 2010).

discussed above (observing, counting, measuring, asking): an activity happening in a variety of methodological contexts. It should also be acknowledged that before students described their method of data analysis as coding, many students were already dealing with large qualitative datasets: they were reading, re-reading and immersing themselves in the data, and also identified key themes and cross-connections in these data. The same goes for the statistical analysis that I described earlier. Many of the earlier dissertations used basic quantitative and statistical analysis in their research, without actively describing the *process* of how they reached their conclusions. This does not imply that methods of data analysis were not as interesting or thought through, but it does indicate that for undergraduate dissertations, the focus might have been more on the final product, the outcome, and less on the presentation of what students actually did and learned along the way.

Mapping landscapes versus modelling landscapes

The longitudinal perspective offers the opportunity to study shifts over time, but also demonstrates some striking *consistencies*. In physical geography, fieldwork methods have remained relatively similar over time, with a focus on active

128 *Exploring the skills of geographers-in-the-making*

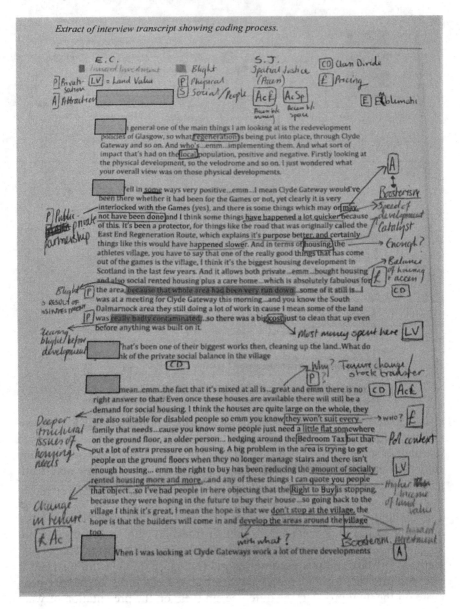

Figure 6.3 Example of coding process (Adamson, 2010).

field encounter and measurement, but innovation in research instruments has arguably made things easier for later students. In terms of data analysis, though, the dissertations in the archive reveal a stronger emphasis on models – either newly developed ones or existing models from the literature – as the key structure of their data analysis. One example is Doug Reid's dissertation *Can published models of coastal beach response to changing wave events be transferred successfully to U.K. inland lake locations?* (1998). Reid's aim was to find out if models that already exist for certain environments can also be *translated* to other geographical scales and contexts. He described this translation of models to other scales as something distinctive to geography:

> The range of scales that the geomorphologist works in is a distinctive part of our discipline, and has been considered a problem in regards to progress in theory development within the subject. But these wildly varying scales can be considered a skeleton of our discipline, and with understanding of the skeleton may come understanding of the rules linking events and forms on different temporal and spatial scales.
> (Reid, 1998: 6)

This approach distinguishes such 1990s physical geography dissertations from their earlier counterparts: in earlier dissertations students focused on one specific case study (and in the 1980s physical geography at Glasgow was seemingly in a state of crisis), with mainly biogeographical projects conducted and barely any geomorphological ones).

One example from 1978 perfectly demonstrates the difference with the 1990s dissertations. In The *glacial geomorphology of the Corrour area* (1978), Ian Young explains why he has chosen this particular focus of inquiry: very little has been written about the Corrour area. The starting point is hence comparable to the tradition of regional geography, merely identifying places as yet little-researched, and not that of 'systematic' geography where one case study might be expected to tell us more about certain kinds of landscapes in general. Young's methods for data collection and data analysis were thorough: a long period of research in the field studying and measuring stone orientations and local relief, followed by the analysis of (existing) aerial photographs. All the academic references and secondary source material referred to aim directly at this area, not to any methodological sources. The result of his research is then a map of suggested ice movement in the area (see Figure 6.4) a very extensive research project.

This map is a folded appendix of the dissertation. When unfolded, it is more than 1 m² in size

The physical geography projects of the 1990s were not entirely different in ambition and execution, although they do often perhaps express a deeper appreciation of more general 'laws'. For many of the projects of 1994 and later, the improvement of research facilities within the universities, such as laboratory facilities for chemical analysis, stream tables, and the flume, combined with the growing expertise and support of academic staff, meant that projects were expected to transcend their own case study, adding not only 'regional-specific' knowledge to the discipline but also more theoretical, conceptual, or methodological insights.

130 *Exploring the skills of geographers-in-the-making*

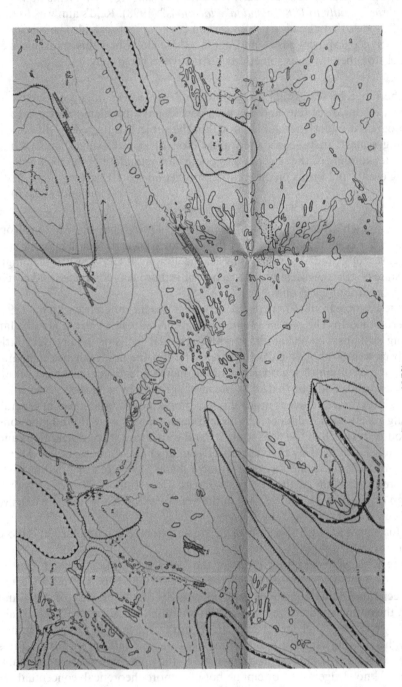

Figure 6.4 A map of suggested ice movement (Young, 1978).

Methods and research design frameworks

Over time, the design of the methods received more explicit attention from students, which is in accordance with the general increase of attention that methodologies received. So, besides describing *what* they were doing, students also discussed *why* they planned to do it that way. In many physical geography dissertations this was already happening since the 1990s, with very specific descriptions of how certain research set-ups were designed and why they were designed that way. The description of methods, such as questionnaires and interviews, were often limited to answering *who* was interviewed, and about *what* that person was interviewed. It was not until quite recently, around 2010 onwards, that many students begin not only to describe who they talk to and why, but also sometimes make an explicit connection back to the overall research aims and objectives, demonstrating how each interview question is included to research a particular aim. There are dissertations that are very much planned and structured from the start, and others which emphasise the research process as more journey-like, open for surprises and detours. However, from the 1990s onwards, and even stronger during the 2000s and early 2010s, these decisions on planning and design are elaborately discussed. From the most recent cohorts studied (different cohorts of the 2010s) the most referred to pieces of methodological literature cited in the dissertations are *Practising Human Geography* (Cloke et al., 2004), and *Methods in Human Geography* (Flowerdew and Martin, 2005). The increase of reflections on methods of data collection and analysis suggests an increase of attention to this in the undergraduate curriculum in Glasgow as well. The consistency of references to specific methodological textbooks[3] indicate that this development happened 'top-down', so was shaped by suggestions and requirements of supervisors and other academic members of staff.

Ethics and positionality

With students becoming more reflective in their research, and simultaneous changes in the rules and guidelines around the doing of research, thinking about research ethics in general, and the relationship between researcher and researched in particular, has changed considerably over time. The biggest shift in the treatment of ethics in the dissertations is perceptible in the dissertations from 2006. Since the mid-2000s, students have been required to gain approval from the departmental Ethics Committee before doing any form of fieldwork which include activities such as speaking to, collaborating with, or studying individual people or groups of people. Since 2010 the ethics formalities have become stricter, with students studying populations deemed 'vulnerable' having to get approval from the Ethics Committee of the College of Science and Engineering, instead of through the committee on departmental level (as is still the case for most dissertations, except for the ones involving 'vulnerable' research subjects). This potentially discouraged students from researching children or those with mental capacity issues, populations that had previously been quite extensively researched in the Glasgow dissertation projects. This particularly influenced the social geography dissertations written in the context of the option course 'Social Geographies of 'Outsiders'': whereas it was still possible to research children's geographies through adult respondents and documentary sources, it was more difficult to get approval for actually *talking to* or *observing* children:

> In order to undertake this research ethically, certain research methods that involved direct contact with children or vulnerable adults were omitted and discourse analysis was employed as an appropriate and valuable research tool.
>
> (Herd, 2014: 14)

This is an example of how the formalities concerning research ethics fundamentally influence the foci of inquiry of students' knowledge productions. Simultaneous to the shift in formal ethical requirements, research ethics became a standard element tackled in most human geography dissertations. The urgency for a stronger emphasis on ethics goes hand-in-hand with more in-depth qualitative research in the booming subdisciplines of social and cultural geography, which in the 2000s relied more on personal input from *individuals* instead of input about groups and more 'factual' circumstances. Whereas some questions concerning research ethics were already addressed in dissertations from the 1990s, in the 2000s the formalities became standardised, and even in dissertations drawing upon interviews about themes that are not very 'personal', for instance on place promotion, include copies of consent forms as well as receiving some comment in the dissertation itself.

Besides these forms of ethics that require consent from participants, signed forms and reflections on the research methods by the student-researcher, students also start to write differently about *their own* role as a researcher. Especially in ethnographic projects, students become more of an 'insider' than an 'objective' outsider. This asks for reflections on one's own position as researcher, as an issue of 'real' engaged ethics during the research rather than just the 'bureaucratic' ethics of securing formal approval to undertake the research. From all sampled dissertations, mentions of research diaries are first found in 1998, but become more common in 2002. In such diaries, which were maybe already kept by students in earlier years but became an integral part of the collected data from the early 2000s, students reflect on their own attitude, feelings, and thoughts. The data in the field diaries became 'evidence' in the dissertation text. Gillian Crawford, researching racial harassment experienced by asylum seekers and refugees in Glasgow, writes in her research diary:

> Beginning to feel very frustrated as it looks like there are less and less possibilities of interviewees – numbers coming to the drop-in have declined and those that come sit with friends and it's too awkward to

broach the subject. I ended up leaving early as there was nothing for me
to do. Research diary 14/8/02.

(Crawford, 2002: 27)

It becomes clear from this citation that the combination of approaching people
and talking about such a sensitive subject is difficult for her. A few pages further she writes:

> I felt my quite [sic] nature was preventing me from speaking with people
> and had to learn to overcome my own insecurities.
>
> (Crawford, 2002: 30)

Crawford worries about how potential participants would think of her. Her
choice to volunteer at the refugee centre while doing this research causes her to
"maneuver [sic] between being an insider and an outsider ... I was aware that I
was entering an already established group and I was firmly on the periphery"
(Crawford, 2002: 30). In the end, she managed to interview five people at the
centre, even though her aim was to interview more.

Although Crawford researched a group that can be seen as vulnerable, she
still struggled with her own role as an outsider and as a shy student, at which
point broader ethical concerns – to do with researching this particular population – intersected with more practical matters of how she negotiated various
different aspects of her own positionality (and indeed personality). There are
many other examples in which these struggles have been prominent, maybe
because of existing differences in power between student-researcher and participants. These differences are often based on age or on gender, as experienced
by Gail Foote in her research for her dissertation *Becoming 'invisible': the hidden spaces of Glasgow's homeless* (2002):

> Although the power relations tended to be in my favour, the homeless did
> have some power in that some intimidated me, which resulted in me, as a
> young woman, feeling vulnerable and choosing not to interview them.
>
> (Foote, 2002: 10)

Other times, the power relations have more to do with the specific 'status' or
authority of the interviewees, as Alasdair Bisset experienced in his dissertation
on the experiences of national identity associated with Scottish army regiments. He writes about adjusting his own appearance in preparation of the
interviews:

> The thoughts of 'power relations' became an important consideration for
> both interviews. I would be interviewing men who held or had held fairly
> senior positions in the Infantry, for example a Major can command
> upwards of one hundred men. With this in mind I decided that it would
> be a more conducive interview if I were clean shaven and dressed smartly.

Fortunately, the atmosphere wasn't too daunting thanks in part to coming from a military background myself and the friendliness of both officers.

(Bisset, 2010: 17–18)

As Bisset describes, his own background played a role in how he personally experienced the fieldwork and interviews. Lots of students research subjects that are close to their own interests or own background, and this involvement with the research may sometimes be a benefit, but at other times a drawback: from being an ardent Celtic football club supporter researching the impact of Celtic Park, the club's home ground, on its immediate neighbourhood, potentially leading the student to bring their own very strong opinions into questions for interviewees (Irvine, 2002), to being someone familiar with certain religious traditions under research, thus being able to ask important follow-up questions (Kakela, 2014). From the 2000s, students have increasingly acknowledged their own positionality, and have actively sought to *use* it in their research and position themselves as knowledgeable, informed 'research tools', unlike some of their predecessors in the 1970s and 1980s, who tried to be 'as objective as possible' on subjects about which they were actually really quite opinionated.

Graphicacy as the geographical skill

Browsing through the geography undergraduate dissertations, the presence of maps in the majority of them is obvious. Maps might be the first association that many people have with geography, and such an assumption seems justified from the evidence in the dissertation cupboard. There are, however, many more examples of graphic material that are found in dissertations. To explore these visuals further, it is interesting to take a side-step to another source, the edited textbook *Geography: An Outline for the Intending Student* (Balchin, 1970), because in this book Balchin provides a chapter presenting a full analysis of the skills that the 'intending' geographer needs. In it, he distinguishes literacy, numeracy,[4] and indeed 'graphicacy'. The latter is described as follows:

Although literacy is the most fundamental method of intellectual communication, graphicacy is the most distinctively geographical form. Without spatial records such as maps, photographs and diagrams, geography would not be geography. Graphicacy is the educated skill that is developed from the visual-spatial ability of intelligence, as distinct from the verbal or numerical abilities.

(Balchin, 1970: 28)

Being skilled in graphicacy supposedly distinguishes the geographer-in-the-making from students in other disciplines. On the next page Balchin explains how geographers' graphicacy relates to the specific skillsets of other academics:

It [graphicacy] is no longer looked upon as an independent art – almost a curiosity – but is recognized as a fundamental support for the whole of geography, distinctive in kind but analogous in function to the fundamental supports of other subjects. Almost every subject has its own special methods of making visible what is really invisible. Thus meteorology depends on instrumentation to illuminate the invisible atmosphere, history upon documents to disinter the obscure past, and economics upon statistics to isolate data that are diffusely concealed among other aspects of daily life.

(Balchin, 1970: 29)

In this following sections, Balchin's threefold classification will be followed: the map, the photograph, and the diagram.

Maps

Many dissertations include one or more maps. However, there is a distinction between maps made as a *research objective* of the student's project, maps displayed to contextualise the research area, or maps displaying results from the research. The first category covers maps that are the result of surveying exercises. In later decades, such exclusively surveying projects are not found in the dissertation archive, but this relates to the earlier-discussed presence of the other undergraduate degree programme that the department offered, Topographic Science, from 1964 through to 2004. Maps, though, remained central for many other geographers, even if their research was not about surveying. Maps as contextualising device have been used from the 1950s through to the present, simply to indicate where the region discussed, the fieldwork location or the case study area is located in the world. Contextual maps can be more than 'just' about location. Such maps were often hand-drawn and copied from other sources, ones rarely explicitly cited in the earlier dissertations, but for the more recent cohorts Google Maps and Google Earth are often used, with no student cartography involved other than occasional additions of symbols or 'pins', and cited.

It becomes more interesting when moving on to the third type of map: maps used to display some of the findings of the research. Visualising questionnaire results on this map is not only useful as a form of visual report in the dissertation, but is also a method for analysing the data. Maps, in all their sorts and types, have been present throughout the decades, and remain a very central element in the works of student-geographers. Advanced mapping skills, however, are mainly, and unsurprisingly, only conspicuously present in the early 'topographic science' projects.

Photographs

Similar to maps, photographs can have different functions: contextualising the research question, displaying the study area, explaining the methodology, or as

136 *Exploring the skills of geographers-in-the-making*

a visual supporting the results or findings. Photographs are, just as maps, found in dissertations produced by all cohorts in this study, but are also influenced by technological developments. An example of a black-and-white Polaroid is found in Figure 6.5: Peter Crabb added several photographs to his dissertation *The Parishes of Tarbolton and Stair* (1958).

The handwritten caption of the photograph demonstrates why he took this photo: it pictures a combination of several elements of the region that he chose to study for his regional dissertation. In it, he not only shows how 'his' region looks, but simultaneously also emphasises some key landscape features 'typical' of this rural region, covering elements such as vegetation and land-use. Although colour photographs appear from the 1970s onwards, black-and-white photographs do not disappear completely until the 1990s. Figure 6.6 displays a photo from public transport taken by Neill Birch for his dissertation

Figure 6.5 Polaroid (Crabb, 1958).

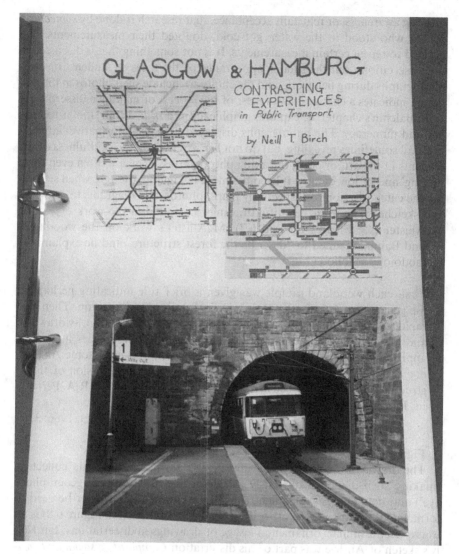

Figure 6.6 Dissertation cover (Birch, 1986).

Glasgow and Hamburg: Contrasting experiences in public transport (Birch, 1986): these images are allowing the reader not to visualise a region, but rather the objects that he was studying.

Whereas the earlier discussed increased attention to a methodology section in the dissertations is perceptible from the 1994, it is also around this time that students actually become present *themselves* in the photographs. Although authors of most physical geography dissertations from these cohorts barely refer to themselves in the first person, this presence in photographs indicates a growing awareness, or reluctant acceptance, that research is done by *someone*: a person who stood in the water, got cold, doubted their measurements, and decided to repeat certain measurements. It is not something that is discussed in the dissertations, but the sudden appearance of the faces of students on such photographs during fieldwork, and the choice to include such photos in the diagrams, indicates a change in awareness of the *process* of making a dissertation.

In Balchin's chapter, he explores graphicacy as a *skill*: a skill similar to literacy and numeracy. The photographs displayed here are informative, interesting, and sometimes revealing, but do not *per se* reflect on the skillfulness of the students in the art, or craft, of taking photographs. They are often even quite 'boring' and of low quality. It becomes more intriguing, though, when stretching the categorisation of respectively maps, photos, and diagrams to drawings and sketches. The most beautiful drawings are found in the work of Robert McAllister (see Figures 6.1 and 6.7). McAllister's work on the woodlands around Balmaha aimed to "describe the forest structure" and he explains his methodology as follows:

> First, each woodland sample was given a brief title indicating perhaps the general type of woodland, its physical setting, and location. Then a rapid, semi-diagrammatic sketch of the woodland structure was drawn and annotated with the names of the most important species in their appropriate strata. ... To give greater precision, and even some small degree of objectivity to the records, a slightly simplified version of the structural formula described in the paper by C.S. Christian and R.A. Perry in 'Journal of Ecology' (see references) was employed.
>
> (McAllister, 1966: 3)

Figure 6.7 displays some of these 'semi-diagrammatic sketches'.

The drawn transects arguably offer flexible forms of both data collection (making such diagrams obviously asked for a well-developed 'geographer's eye') and data analysis, in which regard the drawings would later be used to formulate the findings. The other drawing displayed here (see Figure 6.8) is an example of a similar 'ornamental' usage of drawings in dissertations: Ian Kelly's sketch of Airdrie was part of his dissertation *Comparative social and environmental studies within Airdrie* (1974). In this, he compared two methods of social area analysis by applying it to a case study. The attached marker's report calls it "a rather mediocre piece of geography" with "quite mundane conclusions" (Kelly, 1974: attached marker's report), and although his sketches look pleasing, they also do not offer any additional 'knowledge' to them: it is not clear in this case why a drawing (except, perhaps, using such sketches as a way to display one's own skills) does more than a photo could do.

The inference might further be that for these authors the essentially 'academic' purpose of the dissertation, towards which all its component nowadays should, it seems, ideally press, was less than clear. Yet these drawings, the styles on show and indeed their mere presence, perhaps *do* open a window on a

Figure 6.7 Semi-diagrammatic sketches (McAllister, 1966).

140 *Exploring the skills of geographers-in-the-making*

Figure 6.8 Fieldwork sketches of Airdrie (Kelly, 1974).

different conception of geography to that which now obtains: one imbued with senses of visiting a place, finding ways to describe or evoke that place, and doing something to suggest a long lingering in the place, such as sitting to complete a field sketch.

Diagrams

Balchin distinguished the three classes of maps, photographs, and diagrams, but acknowledged that they overlapped and foresaw even less clear boundaries for the future:

> It is certain that the overlap will continue to increase as geographers seek additional ways of combining the light shed on special relationships by diagrams with the speed and comprehensiveness of photographs and the analytical explicitness of maps.
>
> (Balchin, 1970: 42)

There are indeed many examples of diagrams that are based on photographs or maps and are in some sense 'hybrid' graphs. One example is Jean Burns's diagrams/drawings of Norse house types (see Figure 6.9) in her regional dissertation *A geographical account of the Parish of Wick* (1966).

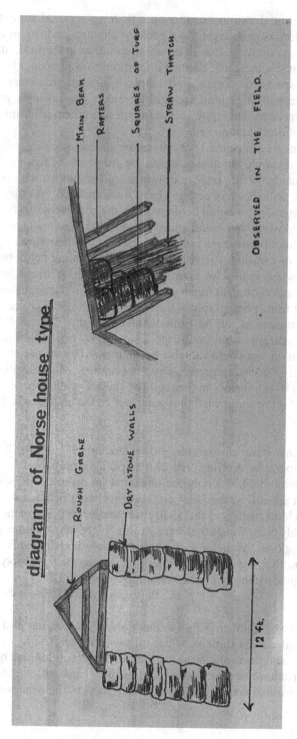

Figure 6.9 Norse house types, observed in the field (Burns, 1966).

It seems like the drawings are very specific for a regional study that also covers descriptions of soil, vegetation, geology, climate, land use, and agriculture, but again the drawings are based on observation in the field, and reflect what the student saw there and deemed important. Such drawings hence highlight specific aspects, describing and emphasising certain elements of the regional landscape of Wick (and, it might be said, if unknowingly, paralleling work on 'architectural geography', housing types and their diffusion as a window on 'folk cultures', and more).

Graphic elements can be used not only as explanation, contextualisation, or justification of research, but also as *validation*. Some of the figures (especially McAllister's, but also other sketches, drawings, and graphs from his contemporaries) reveal a skill that has perhaps been lost, replaced by other skills, or ceasing to be regarded as a skill deployable by the geographer-in-the-making. Making good drawings and useful landscape sketches requires a good 'geographer's eye': taking a photograph is faster and easier, but lacks the analysis as an integral aspect of creating that visual element. Translating field observations into useful drawings is thus an example of being 'literate' in reading a landscape, one that a later student might possess as well, but one arguably less easy to grasp by or from photographs. Of course, certain older projects, such as regional dissertations as well as geomorphological dissertations that aim to describe the landscapes (and landscape histories) of one region or environment, depended on – or could valuably be augmented by the use of – this skill more than is true of many present-day dissertations: visual–spatial literacy might have lost its dominance in the required skillsets geographer-in-the-making over time. Asking about what 'has gone lost' in the undergraduate curriculum, Philo says:

> Probably some basic cartography. Some students … used to be really good at cartography, drawing their own maps. They probably had a much better understanding, not only the ones that go on to do the specialist courses, but actually probably all our students had a richer cartographic tradition behind them and the ability to produce and use maps. There were probably other things that were strong. There was something about a skill in 'regional synthesis' and reading the synthesis in landscapes in front of you that was part of an older geography that probably disappeared to some extent, but not completely.
> (C. Philo, interview, 10 December 2019)

Graphicacy, as well as other skills – numeracy and literacy, but also computational skills and critical thinking – are intertwined with the foci of inquiry, methodologies, and fieldwork locations, and research facilities used by students for their dissertation research; and it is impossible, if not undesirable, to judge the loss or decrease of certain skills without alertness to these many contextual dimensions. Seeing a discipline as being in motion and continuously changing means that necessary skills will also move on.

Conclusion

In many ways, the analysis in this chapter reflects back on the spatial and social contexts as well as the practicalities of fieldwork. These spatial and social contexts, then, are clearly entangled with questions of *how* dissertations are made. The consideration of data collection methods revealed some overlapping practices in very different research projects: observing, counting, measuring, and asking were four of the 'activities' that were, in almost any combination, used in the majority of the projects. There was, however, a variety in the use of different 'instruments' in data collection, and over time the description of why *these* methods and instruments were used increased. Shifts in data analysis methods are mainly influenced by improvements of the university's facilities, as well as by developments and increasing accessibility of data analysis software: however, use of qualitative data analysis software is barely traceable in the dissertation archive. In physical geography, fieldwork methods have remained relatively similar over time, though later dissertations reveal a stronger emphasis on modelling. Shifts in methods were sometimes connected to broader disciplinary changes (such as the emphasis on statistical analysis, and thus the need for methods of data collection permissive of such analyses), but at other times were 'smaller' changes connected to one or two subdisciplines. This account leads to an image of an overall geography discipline that for some (earlier) cohorts was relatively unambiguous, with many students producing relatively similar knowledge productions, but for some (later) cohorts became very broadly stretched, almost evoking questions how *on earth* students could still be part of the same curriculum and department.

Certain methods, or combinations of methods, were strongly related to specific subdisciplinary traditions: for instance, participant observation and ethnography are relatively common in social and cultural geography. This does not mean social or cultural geography dissertations always used such methods of data collection, on the contrary: many of the earlier social geography dissertations are quantitative in nature, and rely on the analysis of demographic data and on questionnaire surveys. Analysing methods in a longitudinal study such as mine, is thus complicated, because the many cohorts of dissertations demonstrate shifts in methods *over time* as well as differences in methods between subdisciplines (in, for instance, one cohort). This analysis reveals what it actually means to be a geographer-in-the-making: what does one *do*, what are reliable sources, and how do student-geographers 'deal with' such sources? Unpacking such questions helps to understand the often overlooked perspective of students on the discipline, and on disciplinary traditions, and disciplinary knowledge in specific.

Intermezzo 6: Commuting

Already from the regional studies of the 1960s and 1970s, commuting or, as it was often called then, 'the journey to work' had become a standard element or theme of students' dissertations. For instance, in their regional dissertation entitled *Muirkirk and Sorn: Hope or Despair?* (1970), J.C. Darroch considered the themes of industry and employment by discussing employment opportunities outside of the parishes that were studied:

> Despite the falling prospects of employment throughout the two parishes, from decline of the coal and textiles economies, very few of the displaced have moved on to the land, but prefer instead to 'journey to work' outside the parishes, where they can obtain the better financial rewards of mining or industry.
>
> (Darroch, 1970: 34)

In the 'systematic' dissertations (specifically urban geography projects) the journey to work has also been prominent, the question of commuting becoming the main emphasis of study instead of 'just' one aspect of recounting the region. Darroch's peer, John Weir, discussed the 'journey to work' in East Kilbride in his dissertation *Towards a further understanding of commuting patterns in the new town of East Kilbride* (1970), which of course also drew in the New Towns angle just reviewed. By means of statistical analysis of data provided by both East Kilbride Development Corporation and nine different companies located in East Kilbride (e.g. Rolls-Royce, Dictaphone Co, Schweppes), Weir probed the benefits and drawbacks of commuting – both for the business and for the region as a whole:

> Whether commuting is desirable or not is quite a hotly contested question. Certainly a very high proportion of commuters travel by car – either their own or a workmate's – e.g. Rolls-Royce have been forced to build two large new car parks in addition to the original two and this must inevitably cause congestion. It has nevertheless been suggested that this is the necessary price to pay for general benefits like segregation of workplace and home giving a certain flexibility.
>
> (Weir, 1970: n.p.)

Commuting patterns and the social dynamics of so-called 'commuter villages' were studied throughout the 1980s and 1990s, but fell out of favour from the year 2000. There are many other questions concerning transport and employment that continued to be researched after that date, but these dissertations emphasise commuting to a lesser extent and focus more on sustainability of transport and levels of accessibility to public transport. An exception is the dissertation by Caitlin King, entitled *Urban commuter cycling: Understanding the cyclist and their environments* (2014), which homes in explicitly on commuting. Again, though, there is a clear trend break with equivalent dissertations from a few decades before, because King emphasised *cycling*, a specific, sustainable mode of transport, but also the cultural aspects of transport in a manner consistent with more recent academic literature. In her research, King compared Hong Kong and Glasgow, deploying observation and interview materials to excavate how cultural and social aspects influence the use of bikes for commuting:

> Urban commuter cycling has been shown through this study to be distinctly affected by predominant cultures, environments and social influences within a city. Hong Kong and Glasgow were found to have differing cultural perceptions of the activity of cycling, and this in turn affected the predominance of urban commuter cycling. These cultural attitudes were shown to be heavily influenced by governmental attitude and policy.
> (King, 2014: 31)

Commuting has thus remained an aspect studied throughout the decades, but one more recently researched increasingly through 'a sustainability lens'.

Notes

1 Such a project would nowadays require a lot of attention in terms of 'risk assessment' and 'mitigation strategies' prior to going to the field – it might well not get approved.
2 Although, increasingly in recent years, some students did online research and never 'left' their homes to do fieldwork: in principle, though, they still return from a 'virtual field' to the department.
3 Textbooks themselves raise many questions about canonicity and disciplinary traditions. Sidaway and Hall (2017) provide a fascinating reflection on this.
4 The chapter on numeracy is also very insightful for any historian of geography. In that chapter, written by S. Gregory, the use of the skill of numeracy is explained as something 'new' to geographers: "The need for geographers to be numerate as well as literate is no new development. ... Over the last ten to fifteen years, however, new trends have appeared in the subject, involving the more systematic and organized application of mathematical reasoning to geographical problems. These trends consist partly of a more sophisticated handling of the traditional type of numerical data, using more complex techniques of analysis and synthesis, and partly of the formalizing of known or assumed geographical relationships into general systems that are often capable of numerical or quantitative evaluation" (Gregory, 1970: 43).

Bibliography

Balchin, W.G.V., 1970. *Geography: An Outline for the Intended Student*. London: Routledge & K. Paul.
Bibby, J.S., Mackney, D., 1969. *Land Use Capability Classification*. Harpenden: Rothamsted Experimental Station.
Bisset, A., 2010. *National identity; The perceived affect of the amalgamation of the Scottish regiments*. Undergraduate Dissertation, University of Glasgow.

Booth, R., 1994. *An investigation into the relationship between beach morphology and beach sediment characteristics, with reference to the loch-shore site of Cashel, Loch Lomond, Scotland.* Undergraduate Dissertation, University of Glasgow.

Cloke, P.J., Cook, I., Crang, P., Goodwin, M.A., Painter, J., Philo, C.P., 2004. *Practising Human Geography.* London: SAGE.

Collison, J., 1974. *A geographical study of organised religion in an urban area.* Undergraduate Dissertation, University of Glasgow.

Crawford, G., 2002. *Racial harassment experience by asylum-seekers and refugees in Glasgow and its impact on daily life.* Undergraduate Dissertation, University of Glasgow.

Darroch, J.C., 1970. *Muirkirk and Sorn: Hope or Despair?* Undergraduate Dissertation, University of Glasgow.

Dickinson, G., 1966. *Mountain vegetation in south-western Rhum.* Undergraduate Dissertation, University of Glasgow.

Dowie, A.J., 1978. *Cognitive mapping and the residential preferences of Drumchapel residents.* Undergraduate Dissertation, University of Glasgow.

Eaglesham, D.G., 1974. *Caithness: A study of a peripheral region.* Undergraduate Dissertation, University of Glasgow.

Flanagan, O., 2006. *"Sense of Community, Place and Well-being": an examination of the culture, identity and geography of the travelling community in the North West of Ireland and the impact this has on their overall health status.* Undergraduate Dissertation, University of Glasgow.

Fleming, G., 2010. *Fear of crime: A case study in Glasgow.* Undergraduate Dissertation, University of Glasgow.

Flowerdew, R., Martin, D., (eds) 2005. *Methods in Human Geography: A Guide for Students Doing a Research Project*, (2nd ed.). Harlow: Pearson.

Foote, G., 2002. *Becoming 'invisible': The hidden spaces and places of Glasgow's homeless.* Undergraduate Dissertation, University of Glasgow.

Graham, J., 1970. *The effects of the Industrial Revolution in the Parish of New Monkland.* Undergraduate Dissertation, University of Glasgow.

Gregory, S., 1970. 'Numeracy', in: Balchin, W.G.V. *Geography: An Outline for the Intending Student.* London: Routledge & K. Paul.

Grieve, I.C., 1970. *A capability classification scheme for marginal land.* Undergraduate Dissertation, University of Glasgow.

Hastings, G., 1990. *Counterurbanisation at the periphery. Case study – Baillieston.* Undergraduate Dissertation, University of Glasgow.

Henderson, S., 2010. *Shenavall Bothy.* Undergraduate Dissertation, University of Glasgow.

Herd, K.A., 2014. *The geographies of child abduction and its governance in the United Kingdom.* Undergraduate Dissertation, University of Glasgow.

Irvine, S., 2002. *The stadium and the people: The impacts of Celtic Park upon the local population.* Undergraduate Dissertation, University of Glasgow.

Kakela, E., 2014. *Somali diaspora in Helsinki: Negotiating gender and religious identities in a secular space.* Undergraduate Dissertation, University of Glasgow.

Keith, G.J., 1998. *The effect of tidal variations on sediment concentrations and fluxes within an intertidal mudflat creek, Chichester Harbour, W. Sussex.* Undergraduate Dissertation, University of Glasgow.

King, C., 2014. *Urban commuter cycling: Understanding the cyclist and their environments.* Undergraduate Dissertation, University of Glasgow.

Lewis, N., 2002. *Horizons of the future: Investigation into residents' attitudes concerning the visible existence and visible proximity of Beinn an Tuirc windfarm*. Undergraduate Dissertation, University of Glasgow.

McAllister, R.J., 1966. *Structure in the woodlands around Balmaha, Loch Lomond*. Undergraduate Dissertation, University of Glasgow.

Milligan, C., 2002. *Foreign vodka, designer labels and pricey clubs threaten 'Geordie' culture...almost!* Undergraduate Dissertation, University of Glasgow.

Parker, M., 1962. *A geographical study of the Glasgow Underground Railway*. Undergraduate Dissertation, University of Glasgow.

Reid, D.H., 1998. *Can published models of coastal beach response to changing wave events be transferred successfully to U.K. inland lake locations?* Undergraduate Dissertation, University of Glasgow.

Thomson, C.C., 1966. *Analysis of the shopping facilities of Grangemouth*. Undergraduate Dissertation, University of Glasgow.

Weir, J.C., 1970. *Towards a further understanding of commuting patterns in the new town of East Kilbride*. Undergraduate Dissertation, University of Glasgow.

Welch, C.A., 1974. *Fjaerland I Sogn*. Undergraduate Dissertation, University of Glasgow.

Wilbur, A., 2002. *The Faslane Peace Camp: Mobility and permanence in sites of resistance*. Undergraduate Dissertation, University of Glasgow.

Young, I.H., 1978. *The glacial geomorphology of the Corrour area*. Undergraduate Dissertation, University of Glasgow.

7 Reflections on student journeys into geography

The empirical heart of this book was one collection of undergraduate dissertations, located in a cupboard in geography's accommodation in the 'East Quad' at the University of Glasgow. This collection, comprising dissertations from the early 1950s to the present day, can be researched as a singular source: as a resource for narrating the history of one discipline, of one undergraduate curriculum, and at one institution. Simultaneously, the collection contains a plurality of over 2,600 dissertations, each revealing a different side of geography, a different view on the experience of doing independent research as a student, and a different reflection on being a geographer-in-the-making in a certain place and time. The object-oriented approach taken in this book was aimed at acknowledging the value of the archival collection of dissertations in both its singular and plural manifestations.

Narrating the history of geography from a student perspective started by exploring existing historiographical narratives, from 'usual suspects' such as Livingstone (1992) and Johnston and Sidaway (1979, 2004, 2016) to complementing or competing narratives written by geographers such as Maddrell (2009) and Blunt and Wills (2000). Two central questions in considering this variety of narratives are: first, whose 'voices' should be included, and hence who actually 'makes' the discipline? And second, what is the structuring device in creating these narratives? Is it about geographical *ideas*, about *practices*, about *texts*, or about certain supposedly pivotal *individual contributions*? Engaging with canonical historiographies as well as other narratives demonstrated a clear gap in the voices that were staged. Whereas professional, established academics were, evidently, central in many narratives, and while some historiographies emphasise geographical knowledge and ideas arising outside of academia, the group of student-geographers – in numbers easily the largest group of people found in universities – are often only present as *consumers* of geographical knowledge (for instance, as a target audience for textbooks: Sidaway and Hall, 2018) and almost never as *producers* of such knowledge. Students being the main 'practitioners' of geography vastly outnumber professional academics, and they not only consume knowledge but also reproduce it (Philo, 1998), and just sometimes they too produce new knowledge. Researching student knowledge productions from the past provides a unique, bottom-up

DOI: 10.4324/9781003348139-7

perspective on how geography has changed over time: becoming-a-geographer is a process (Lorimer, 2003), and this process takes place on a 'middle level', in part set between 'elite' academic geography and grassroots versions of geography (Philo, 1998).

Studying dissertations from one discipline, one undergraduate curriculum, and from one department might not be indicative of the experiences of *all* geography students in Scotland, in the UK, or globally, but this local scale of inquiry is still very valuable: knowledge is *always* situated, and approaching science as a 'view from nowhere' (Haraway, 1991) would ignore the social processes integral to the production of knowledge and only emphasise the contributions of those established, well-known, and 'elite' individuals. Exploring the dissertation archive, then, addresses many 'micro-histories' in a specific social and educational context. Although the focus was primarily on what might be deemed finished work, the written dissertations themselves, this book also included inquiries into wider networks of which students were a part and into the dissertation research process as a whole, including doubts and feelings that students experienced during their first independent research projects. Acknowledging, thereby, that knowledge is not something 'placeless' means that one dissertation collection at one institution *is* potentially a highly instructive starting point for this disciplinary history from below, representing many small voices of the discipline and, in this, demonstrating that geographers-in-the-making play a fundamental part in the history of geography (as themselves minor 'makers' of that geography).

Disciplinary histories from below

Dissertations are not entirely 'free forms' of writing but are distinguishable from other forms of formal assessment such as exams. The educational contexts in which dissertations are written strongly influence the contents of these dissertations: specific courses, facilities, and individual staff members affect the choices that students make. This raises the question of *who* exactly produces the dissertations: to what extent are student-geographers independent and how much are they influenced by the networks within their department? Multiple students reflect on the influence a specific member of staff held in their choices concerning their dissertation research; and even when it is not about a specific supervisor (sometimes called an advisor, implying a more 'hands-off' role), the curriculum presents a certain interpretation, or version, of geography, implying that a geography student in Glasgow will have a slightly (or perhaps significantly) different education from a geography student in, for instance, Edinburgh, London, Amsterdam, or anywhere else. This educational context and the local variations of geography seem to provide less 'ownership' of students on their research, but the same might be said for 'established' geographers: they are undoubtedly influenced by their intellectual backgrounds and by the institutions they studied and work at, and so the answer to this question of *who* produces the knowledge is always destined to be rather diffuse. To some extent, in dissertations, it is less covert than in other 'publications' because the documents themselves very much 'signal' these networks by mentioning of the supervisor, naming the institution on the front page, and via the acknowledgements.

This research aimed to stage the voices of many geographers who are often neglected. Undergraduate dissertations are usually seen as ephemeral sources, read by only a handful of people, and yet writing a dissertation is a shared, formative experience for many, if not all, established academics, and thus a very recognisable practice. The centrality of undergraduate degrees in university departments (in terms of centrality to curricula, how they 'gather' together so much conceptual, methodological, and (sub)disciplinary learning and the substantial workload of staff supervision) as well as in the personal reflections of past students on their 'student life' (in terms of the undergraduate degree[1] taking up several formative years whereas postgraduate degrees are often either shorter – for instance, 1-year research Masters – or only experienced by a select group – for instance, PhD degrees) arguably demands a similar centrality of their voices in the discipline's history and practice. This does not mean, for instance, that academic journals and conferences should suddenly start carrying and presenting large quantities of undergraduate work, although maybe there could be a commitment to occasionally publish versions of high-quality undergraduate dissertations or showcasing such work in other ways. Certainly, though, it should mean that the voices of student-geographers not be kept silent solely *because* they are the voices of relatively inexperienced geographers. This book aims to, even given the reality of working at institutions with massive student numbers and an enormous workload, stimulate academic staff supervising, teaching, and examining undergraduate students to keep recognising the surprising, beautiful, meticulous, well-written, exemplary, fascinating, and/or critical works of research produced by at least some of their students.

The lived experiences of geography students 'entering' the disciplinary community

The in-depth analysis of and engagement with many small knowledge productions, produced by so many different students, also shines a light on the sociology and geography of how knowledge is produced: the realities of questions about access, money, travel opportunity, language, and social, intellectual, and practical support unavoidably, and sometimes profoundly, shape the intellectual knowledge productions of these geographers-in-the-making. The shift from research in, or close to, Glasgow to more international research is clearly perceptible in the archival collection, and there are multiple probable causes: generally, it has become cheaper and more 'normal' to travel abroad; student mobility programmes such as ERASMUS have led to new possibilities for students to go abroad, and the student population at the University of Glasgow has mutated from a population dominated by Glaswegian students and students from the West of Scotland's Central Belt to comprising a more diverse and international group of students. Despite this continuing internationalisation of the dissertation research for the entire time period studied in this research (from the 1950s to the present), the majority of the projects were still Scotland based. International projects bring specific practical challenges to do with accessibility, costs, and language, but many other challenges are independent of fieldwork location: for instance, the influence of the weather, use and availability of instruments, and difficulties in finding participants. Some of these issues might be dealt with by determined effort as well as support from strangers, but others are fixed by reliance on social networks and a stronger personal (and familial) socio-economic position. Already, before (academic) careers actually start, some geographers have enjoyed fewer opportunities and less time to focus on study, potentially a very crucial factor in later career possibilities. The privileges of having a strong support network shape the academic possibilities of and for students, and, although local or regional research is by no means 'lesser' than international research, many students who have travelled to, for instance, Tanzania, Iceland, or Norway, likely on expeditions associated with particular staff members, reflected on the positive influence for themselves, both as a geography student and as a person. Especially, in terms of future relationships between the UK and the EU, it will be necessary to consider the specific challenges that some students will likely experience more than others in their choices concerning fieldwork options.

Every single dissertation is a 'small story' of a student travelling to exciting or not-so-exciting places, of helpful or not-so-helpful supervisors, parents, and friends, of insightful or not-so-insightful reflections on what it meant to be a student, a geography student, at the University Glasgow; but every single dissertation also constitutes an intellectual piece of geography, written in a specific social, spatial, and educational context. In comparison to other writing on the intersection of the history of academic education and the sociology of knowledge production, the emphasis placed on dissertations as advanced here goes beyond the common treatment of students as relatively passive knowledge consumers. This emphasis has included reflections on the responsibility and autonomy of students in shaping their own research projects, organising their travels to fieldwork locations and building positive student–supervisor relationships. This book explored these spatial and social contexts of undergraduate students in depth and, by combining a historical perspective with this sociological and geographical perspective, has given agency to this under-represented group within the outer reaches of the academic community. By acknowledging their importance, both in size of the group of students and in their produced knowledge, their experiences deserve to be studied just as the research experiences of 'established' researchers are studied by sociologists of science.

The decline of regional geography, extensively discussed in this book, coincided with an increase in the variety of 'scales' of research and a broadening conception of what 'the field' actually is or could be, notably in how it is no longer just a clearly delimitable 'region' such as a Scottish parish or county. Questions of scale are obviously a spatial question, but are also connected to both specific foci of inquiry and practical methodologies: for instance, the possibilities of doing virtual fieldwork have gone hand-in-hand with the appearance of online surveys and social media analyses. With present-day cohorts of students and graduates with exceptional experiences in virtual research because of the COVID-19 pandemic, it is expected that this move to 'the virtual field' might have an impact on the nature of dissertation research going forward (both in how virtual research may have inspired staff and students or, perhaps a contrary development, with very strong enthusiasm reasserting itself for 'outdoor', 'real life' fieldwork as an 'antidote' for the experienced lockdowns). Expansions of fields are thus instigated by practical circumstances, as well as by disciplinary trends and innovations. Combined, this mix fuels a broad variety of dissertation research experiences, a changing conception of what 'geographical research' actually looks like, and a changing sense of what becoming a geographer actually means. Such observations are inextricably connected to the changing foci of inquiry, conceptual frameworks, and conceptions of what geography is and should be.

Dissertations as sources

Although probably not surprising to many, the relationship between human and physical geography in the curriculum is perceptible in all eras and cohorts. Geography, as a discipline with roots in and connections with multiple conceptual and methodological traditions, is a versatile discipline. Generally, the foci of inquiry such as disaster studies and environmental studies cause convergent motions, drawing together human and physical geography, whereas epistemological differences – concerning what sort of knowledge is to be produced, and how – at times drive them apart. These developments are strongly embedded in the curriculum and are not independent of changing conceptions of geography as a wider academic discipline, but their translation into curricula then decide the fast or slow integration of new concepts, theories, and ideas in the education of student geographers. Becoming a geographer, then, is slightly different from becoming a *human* geographer or a *physical* geographer, although many dissertations do fit well in one of these two categories; the examples that are more hybrid demonstrate that the nature of the curriculum still, to some extent, defies this binary categorisation.

The number of dissertations per subdiscipline shows slow and subtle changes, whereas in some cases it becomes apparent that subdisciplines became popular within the timeframe of a few years. As discussed, some of these timelines relate to wider trends in the geographical academy, while others are a combination of such disciplinary developments and the influence of individual members of staff. It can be presumed that, at least to some extent, departments and hiring committees, when deciding on recruitment strategies, specifically identify a lack of teaching expertise on a certain subject or a specific expertise that is not 'covered' yet by the existing staff. This is not to ignore entirely the innovations and new ideas that academics can develop while already working at the university, but it cannot be denied that the expertise of newly hired staff provides a more *instant* change. Strong impacts are also found in new honours options, such as the courses on Natural Hazards and the Social Geographies of Outsiders. The value of quantifying a large collection of dissertations over a long period of time lies in its potential to identify both slower and faster shifts over time and to compare these shifts to ones displayed in existing timelines that are written down about the history of geography. It is a tool to see the interplay between geography in its research and its teaching contexts: a starting point to address disciplinary, institutional, and educational relationships between research and academic education.

Narrating the history of geography based on these sources from many geographers-in-the-making offers a kind of 'youthful' perspective on geography: students write dissertations they deem geographical, influenced by what they have learned along the way. This youthfulness – not in terms of age *per se*, but in terms of phase in their academic education and career – leads to perspectives on the discipline that inherently will be framed a touch different than the perspective of geographers working in the discipline for several decades. As mentioned before, *of course* students are influenced by the professional geographers in their department, but among the hundreds of dissertations, there are quite a few that address a very original perspective on the discipline and on their own roles as geographers. Not every work is a slice of innovative, brilliant knowledge production, to be sure, and that is not what the aim should be. However, every dissertation is a unique, individual interpretation of a geographical question or theme, and should be acknowledged as such.

Although many projects are very different in their focus of inquiry, even between older and newer examples there is still a striking overlap in certain core 'verbs' of data collection: being, observing, counting, measuring, and asking. Generally, however, the diversity *within* cohorts became broadly stretched over time, encompassing an ever-widening range of methods, whereas in some of the earlier cohorts many students produced relatively similar pieces of work using much the same relatively narrow suite of activities (not even necessarily conceived as 'methods'). Shifts in data analysis of course connected across to changing research questions and conceptual frameworks, but were also strongly influenced by technological and IT developments as well as access to laboratory and IT facilities. Shifts in the methods for data collection and data analysis caused the loss and gain of certain skills among the cohorts, which becomes apparent in elements such as maps drawn by students. The considerations on ethics and positionality also demonstrate how, unsurprisingly, students do not always see themselves as fully-fledged geographers, and sometimes struggle with guilt about taking up peoples' time and space in doing their research. This carefulness, sometimes even humbleness, is a good reminder that researchers *are* often supported by many others, from their own network and from networks of strangers. Formalisation of the research ethics process of course emphasises this sort of concern, while also potentially making some forms of research beyond reach for undergraduate students.

The binding and appendices of dissertations have changed so much over time: from folders to ring-binders; from including floppy disks and CD-ROMs to USB-sticks and then online submissions; from appendices such as complete Ordnance Survey Maps to advertisements and articles from magazines and newspapers. The forms of presentation have changed greatly too: from hand-written to typed to word-processed; from hand-drawn maps and glued-in polaroid photos to Google Earth Maps and scanned-in photographs. Indeed, the material dimension of the dissertations might have deserved an entire research project dedicated to it alone. The dissertations as material sources also include many markers' reports in them, accidentally or purposefully attached to the dissertations. These documents are informative, and they yield insights into what was expected of students, how feedback was framed and written down, and how certain grades were justified by the supervisors and other markers. The shift, for instance, from one supervisor being the marker to a system of the supervisor and a second marker marking a dissertation before deciding on a grade, with a third person stepping in if the two initial gradings were too far apart for a compromise to be agreed, is demonstrated clearly by some of these markers' reports.

The dissertations as material sources clearly manifest the labour of producing the final dissertation. Whereas for very recent cohorts there are sometimes multiple copies of the dissertation in the archive, this is never the case for earlier dissertations: of course, the ease of just printing an 'extra' copy for the second marker is very different from handwriting a second version of one's work. To have the dissertations actually in your hand, as the researcher, offers a different experience than just reading the content, and it allows a richer sensibility than would arguably arise simply by reading 'just' the texts in the dissertations. Seeing the dissertations organised collectively in one cupboard really is something different from unzipping a .zip file containing all dissertations of a cohort (as is the case when studying Glasgow's geography dissertations from academic year 2017–2018 or later): the lived experience of doing research, writing up a dissertation, handing in the dissertation, and waiting for it to be marked becomes much more palpable when browsing the crowded stacks of the cupboard.

The dissertations can thus be studied as intellectual sources, adding new perspectives, confirming, rejecting, or reshaping existing geographical ideas. The social and intellectual roles that students play in a department are significant as are the social networks and connections that 'take place' beyond the words written down in the dissertation. The significant experience of writing a dissertation, undertaken by so many, also makes the dissertation a historical source that opens a window on an important rite of passage: annually done by so many, almost a ritualistic movement from being 'just' a student-geographer to being, as it were, an approved, acknowledged, even 'certified' geographer with the degree and academic title as 'proof'. These rituals shift over time but are nevertheless still very much recognisable as this shared experience across time. This multidimensional value of undergraduate dissertations is examined in this book by exploring all these sides: sometimes focussing on the intellectual perspective, other times on the social or cultural perspective, but often acknowledging the multilayered, overlapping use of the dissertations (as singular and plural source materials) for historical research.

There are also pedagogical implications with regard to the usage of undergraduate dissertations as historical sources. Especially by treating the dissertations as windows for the social and cultural 'coming-to-being' of the final intellectual knowledge productions, it becomes evident that many dissertation projects are influenced by social networks, economic support networks, and the related time spent on part-time jobs (and time thus not available for research). The archive reveals a shift to more research occurring away from Glasgow over time, but it is also important to note that the inequality of means *within* cohorts is not decreasing – perhaps even on the contrary. The acknowledgements written by students from more recent cohorts reveal the opportunities some have enjoyed, as well as the resilience of many students to combine external as well as self-imposed obligations with their dissertation research. The impact of such combined responsibilities is not always negative and might lead to innovative perspectives on, for instance, childcare, employment, and student life, but this reality might ask for awareness, or perhaps different forms of guidance, from academics supporting student-geographers in their dissertation endeavours. The pedagogical implications of changes in the social and cultural experience of becoming a geographer over time, alongside the differences between individual students, suggest in some cases the need, for instance, for flexibility, information about funding opportunities, and introductions to a wider network. It also means that some students might already bring very specialised personal or professional experiences with them, which might encourage and motivate students to choose particular foci of inquiry, methods, or fieldwork locations.

Closing the cupboard door

So much knowledge is developed, created, and written down by students, and often these knowledge productions get lost quite soon after the dissertation is handed in, graded, and archived. That said, former students take this formative experience of doing the research, as well as bits of that knowledge, with them into their subsequent lives and careers. The knowledge is hence not lost entirely for everyone, and with certain regards, it is still very much 'out there' in the larger world, but this knowledge is arguably worth being better curated and, perhaps, made more widely accessible. Many professional geographers working at universities will teach, advise, supervise, and examine undergraduate geography students and will at times be surprised – positively or perhaps sometimes otherwise – by their ideas, methods, and analyses. Just possibly, these professional geographers would welcome being able to have ready access to a properly catalogued dissertation archive such as is furnished by the Glasgow collection. Big questions remain, however, about how this archive might be maintained in the future – particularly now that the physical store needs to be supplemented by fresh efforts to update the digital store – as well as curated, rendered widely searchable, and dissertations made accessible as a hard or virtual copy. Likely, the dissertations will remain in their cupboard, dependent on the interest and concern of certain staff members, likely to 'battle' at some point for the need for more physical space.

In this longitudinal research, the increase in the possibilities of international travels, proved dramatically to reshape the experiences of undergraduate geography students at the University of Glasgow, and the same might be said about myself. I sincerely hope that I achieved a very small change in how knowledge productions of students, from whatever level or university, are considered for what they are: sometimes valuable, sometimes crappy productions of knowledge, with an entire 'world' going on behind the words that are written down.

Note

1 Work for an undergraduate dissertation at Glasgow has commonly straddled two academic years, often lasting more than 12 months from initial conception to final submission, with different 'milestones' along the way.

Bibliography

Blunt, A., Wills, J., 2000. *Dissident Geographies: An Introduction to Radical Ideas and Practice*. Harlow, Longman.

Flowerdew, R., Martin, D., (eds) 2005. *Methods in Human Geography: A Guide for Students Doing a Research Project*, (2nd ed.). Harlow: Pearson.

Haraway, D.J., 1991. *Simians, Cyborgs and Women: The Reinvention of Nature*. London: Free Association.

Johnston, R.J., 1979. *Geography and Geographers: Anglo-American Human Geography Since 1945*. London: Arnold.

Johnston, R.J., Sidaway, J.D., 2004. *Geography & Geographers: Anglo-American Human Geography Since 1945* (6th ed.). London: Arnold.

Johnston, R.J., Sidaway, J.D., 2016. *Geography and Geographers: Anglo-American Human Geography Since 1945* (7th ed.), London: Arnold.

Livingstone, D.N., 1992. *The Geographical Tradition: Episodes in the History of a Contested Enterprise*. Oxford: Blackwell.

Lorimer, H., 2003. 'The geographical field course as active archive', *Cultural Geographies*, 10(3), 278–308.

Maddrell, A., 2009. *Complex Locations: Women's Geographical Work in the UK, 1850–1970*. Chichester: Wiley-Blackwell.

Philo, C., 1998. 'Reading *Drumlin*: academic geography and a student geographical magazine'. *Progress in Human Geography*, 22(3), 344–367.

Sidaway, J., Hall, T., 2018. 'Geography textbooks, pedagogy and disciplinary traditions'. *Area*, 50(1), 34–42.

Wilbur, A., 2002. *The Faslane Peace Camp: Mobility and permanence in sites of resistance*. Undergraduate Dissertation, University of Glasgow.

Index

Pages in *italics* refer to figures and pages in **bold** refer to tables.

Academic networks 23
Acknowledgements 23–24, 53
Archive 29, 45, 51
Assessment 20–21, 25
Authorship 28–29

Benchmarking 16
Bibliographies 2
Biogeography 52, 98
Biography 41–43
Bodies 42, 81
Botany 91–92
Bothies 41–42

Canada 40, **40**
Canonicity 2–3
Cartography 65
Channel Islands 39
Climatology 27
Coastal geography 69, 72
Coding 125–127, *127*
Cohortness 30
Commuting 144–145
Conservation 32, 64, 66
Covid-19 152
Cultural geography 41–42, 75–82
Curriculum 16, 26–28, 30, 62

Data analysis 43, 123–129, *130*
Data collection 43, 113–122
Departments 24
Development geography 55, 104
Disciplinary awareness 87–91
Disciplinary community 150–151
Disciplinary identity 6–8
Disciplinary traditions 62, 88–90
Dissertation record card 25–26

Dissertation supervisor 22–23, 25–26
Dissertations as sources 4–6, 61, 150–154
DIY 49

Economic geography 26–27
Egypt 38, 40, **40**, 104
Embodiment 42, 81
England 39
Environmental geography 64–66
ERASMUS 9, 37, 150
Ethics 131–134
Expeditions 39–40, 52–54
Experimental research 44, 72

Feminist geography 103–104
Field 35, 43, 55, *47*, 151–152
Fieldwork 2, 26, 46–52, 113–122
Fluvial geography 44, 72–73
Focus groups 121–122
France 40, **40**

Gender 51, 103–104
Geohumanities 82
Geology 91
Geomorphology 38, 69–74, *71*
Glacial geography 71
Globalisation 9, 36
Graphicacy 134–135

Hazard studies 66–67
Health 107–109
Higher education governance 16–17
Historical geography 27–28, 45
Housing 11–14
Humanistic geography 102–103
Humanities 81
Hybridity 6–8

Index

Iceland 37–38, 40, **40**
Instruments 49, 118
Interdisciplinarity 6
Internationalisation 37
Interviews 48, 51, 121
Ireland 40, **40**

Knowledge consumption 3, 148
Knowledge production 3–5, 148–149

Laboratory 35, 43–44
Language 27, 50, 89
Library 45, 51
Loch Lomond 31

Maps 135
Marxist geography 100–102
Master's Thesis 20
Measurements 117–118
Medical geography 107–109
Methodology 38, 88–89, 113–129, *130*
Microscale 41
Mobility 37, 40, 150
Modelling 127–129, *130*

New towns 11–13
New Zealand 40, **40**
Northern Ireland 39
Norway 40, **40**
NRT 104–105

Observation 113–117
Originality 89
Outdoor recreation 31
Outsiders 44, 79–81

Participant observation 51
Peers 24, 30, 50
PhD Thesis 19
Photographs 135–138, *139*
Physical geography 43, 62–67, 75
Political geography 28, 38, 69
Positionality 131–134
Postcolonial geography 103–104
Power relations 5, 29
Psychology 91

Quality assurance 16
Quantitative analysis 99–100, 123–125
Questionnaires 49, 119–120

Readership 29
Recreation 31
Regional geography 27, **27**, 92, 97–99, 106, 151–152
Research assistants 50–51
Research facilities 24, 44–45, 73, 153
Rural geography 57

Scale 35, 38, 41, 151–152
Scotland 37–39, **37**
Social geography 75–82
Social justice 84–85
Spaces of knowledge production 35–36, 45
Spatial science 99–100
Statistical analysis 123–125
Student mobility 37, 40, 150
Student-supervisor relationships 22–23
Subdisciplines 67–70
Supervision 22–23, 25–26
Supporting staff 24
Sustainability 65
Switzerland 40, **40**
Systematic geography 28, 98

Tanzania 38, 40, **40**, 53, 104
Teaching 36, 78
Textbooks 3
Transport geography 144–145
Travel 36, 151
Troubles 39

Undergraduate research 17–18
United Kingdom 17, 19, 37–38, **37**
Urban geography 28, 38
USA 17, 40, **40**

Virtual world 38, 155

Wales 41
Weather 51, 152

Printed in the United States
by Baker & Taylor Publisher Services

Printed in the United States
by Baker & Taylor Publisher Services